Household Modern Energy Financing

Assessing Photovoltaic Solar Energy Financing Models and Sustainable Energy Transition in Ngaciuma-Kinyaritha Sub-Catchment, Kenya

BY

Amos Yesutanbul Nkpeebo

Research Group on Clean Development

ReGCDEV, Ghana

Email: Aynexe(at)Gmail.Com

DEDICATION

I dedicate this thesis to my beloved benefactors and benefactresses, my mentors and mentees and my supportive family.

ABSTRACT

Solar energy is deemed the single energy resource that is continuously decreasing in price (by 75%), increasing in utility and could effectively contribute to sustainable watershed management. In Kenya, there is an observed acceptance of Photovoltaic (PV) Solar Home Systems (SHS) as the best-fit form of energy to meet rural energy demand. This could potentially displace rural predisposition to woodfuels and paraffin and cumulatively, reduce environmental vulnerability. Photovoltaic SHS are nonetheless challenged mainly by the initial capital cost disbursement, globally and in Ngaciuma-Kinyaritha sub-catchment, Kenya. The focus of this study was to evaluate the economic and environmental significance of different solar energy financing models. It also aimed to analyse different scenarios in order to determine the most cost effective, most sustainable and best-fit financing models that together overcome the capital up-front of solar energy accessing in Ngaciuma-Kinyaritha sub-catchment. In achieving the stated objective, the study adopted and adapted interplay of quantitative and qualitative tools of data collection and methods of data analysis to establish the relationship between energy use and environmental degradation and the ability of households to transition into solar energy use. The study made use of empirical survey instruments including: structured questionnaires (100 cases - using households as a sampling unit), Focused Group Discussions (in the upper, middle and lower zones of the sub-catchment) and interview guides for three selected institutions. The analysis of data followed an objective-based approach in order to emphasize the field observation under each key objective. In analysing the field data, the study made use of simple analytical tools comprising: descriptive statistics (frequencies, percentages and means), chi square analysis, cost benefit analysis, PESTELI analysis, scenario analysis and data triangulation. Using cost benefit analysis, the study perceived a payback period of 6years for SHS that are 50watts and a life payback of 8,685Ksh. It also recorded a payback period of 8years for systems that range between 200wats and 1kilowatts and a life payback of about 100,000Ksh. On the average, the ability to pay for SHS under the cash sale financing model, third party credits model and the solar developer model was observed to be 200Ksh. Using scenario analysis, the study indicates that willingness and ability to pay for multiple utility SHS under the solar developer model is relatively higher than the cash sale model which is the observed status quo in the study area. In a PESTELI analysis, the study perceived a tenable potential in rural energy supply and recommended that an energy user remodelling should be undertaken by the Kenya's energy ministry to foster energy transformation in Ngaciuma-Kinyaritha sub-catchment. It also recommended that the

Ministry of energy (MOE) in collaboration with Global Village Energy Partnership (GVEP) could pilot alternative capacity building scenarios in rural energy use. Finally it gives specific strategies which could be used to improve upon sustainable energy use in Ngaciuma-Kinyaritha Sub-catchment and across the developing world.

ACRONYMS AND ABBREVIATIONS

AEI	Alternative Energy Institute
ATP	Ability To Pay
B/C	Benefit per Cost
BCR	Benefit Cost Ratio
CDM	Clean Development Mechanisms
DAAD	Deutscher Akademischer Austausch Dienst (German Academic Exchange Service)
DNA	Designated National Authority
ETM	Energy Transition Model
FGD(s)	Focused Group Discussion(s)
GEF	Global Energy Facility
GHG	Green House Gas
GVEP	Global Village Energy Partnership
H.E.P	Hydro Electric Power
IAP	Integrated Assessment and Planning
IWRM	Integrated Water Resource Management
LPG	Liquefied Petroleum Gas

Ksh	Kenya Shillings
KPLC	Kenya Power and Lighting Company
MOE	Ministry of Energy
MKEPP	Mount Kenya East Pilot Project
MDG	Millennium Development Goals
NCSA	National Capacity Needs Self-Assessment
NEMA	National Environmental Management Authority
NGOs	Non-Governmental Organization
NPV	Net Present Value
PV	Photovoltaic
PCE	Perceived Consumer Effect
PESTELI	Political, Economic, Social, Technological, Environmental, Legal and Institutional
RETs	Renewable Energy Technologies
SACCO	Savings and Credit Co-operative
SHS	Solar Home Systems
SPSS	Statistical Package for Social Sciences
UNEP	United Nations Environmental Program
VC	Venture Capitalist
WEDI	Women Enterprise Development Institute
WRUA	Water Resource User Association

TABLE OF CONTENT

LIST OF TABLES

LIST OF FIGURES

LIST OF PLATES

LIST OF APPENDICES

CHAPTER ONE

INTRODUCTION

1.1 Background to the Study Problem

Environmental economics was born with the idea of internalizing environmental externalities for ecosystem sustainability (Pigou, 1938; Baumol, and Oates, 1988; Chang *et al.*, 2009). Hitherto, several studies show that the incessant abstraction of primary resources in watersheds often fails to account for the negative environmental externalities that advance the course of watershed degradation (Stern, 2007; Reddy, 2008). This increasing degradation of watersheds resulting from increasing demand for primary resources has resulted into a global refocusing on sustainable use of natural resources making the Integrated Watershed Management (IWM) a paramount concept in the multipronged development approach enveloped in the United Nations Millennium Development Goals (MDGs) (Gilg *et al.,* 2005; Obando, 2005; Owusu 2010).

This study settles on the new curve that distinguishes Integrated Watershed Management (IWM) from Integrated Water Resource Management (IWRM). As a concept, the IWM underlines the judicious use of all watershed resources (renewable and non-renewable) from a holistic point of view without compromising with environmental demands or the ability of future generations to meet their development needs making it imperative to correlate resource use with environmental impacts (GWP 2005).

Among the renewable and non renewable watershed resources subsumed in the IWM approach, energy resources are viewed as the key that unlocks all other development potentials and will continue to be the answer to livelihood advancement in both developed and developing countries (Owusu 2010). Despite this fact, energy use often leaves extensive negative externalities on the global environment. In Africa, the debate has about been the overreliance on woodfuels; 56% and 92% of wood harvested each year is used directly as fuel globally and in Africa respectively (Sankar, 2005; Murray *et al.*, 2009; Carraro and Massetti 2011).

Amidst the apparent negative environmental effect of woodfuel use and its concomitant economic value for about 70% of households in Kenya, there remains general "paucity and inadequacy of information" regarding the quantities and rates of woodfuel abstraction within various catchments (Mahiri, 2003; Kristoferson, 2011). In Ngaciuma-Kinyaritha sub-catchment alone, about 80% of households depend on wood resources and paraffin to meet their basic energy needs (DAAD, 2007, Imenti North District Development Plan, 2008). Given the increasing scarcity of wood resources and the continuous price increase in petroleum products in Kenya, it is possible that this population (80%) could face an energy stress. This austerity questions the sustainability of the roadmap towards achieving a balance of energy mix that fosters sustainable development at the household level.

Solar energy is by far a part of the solution to combat climate change. However, Photovoltaic (PV) Solar home systems (SHS), which since the 1980s, has proven a potential solution to the overreliance on woodfuels and paraffin is still challenged by systemic inefficiencies. Jacobson (2006) and Owusu (2010) illustrate that, owing to the varying household capacities to meet the capital up-front, solar energy is seen to have relatively poor demand and resultantly, less economic and environmental significance. Given the market driven nature of the solar energy use in Kenya, Gueye *et al.*, (2004) sketches out the following models as the major means by which households meet the initial capital cost of solar power systems: (a) Cash Sales, (b) Dealer Credit, (d) End-User Credit, (e) Hire-Purchase and (f) Solar Developer. In the context of these financing models which dictate the product capacities, solar energy application is apparently restricted to household lighting and a few heating processes. This research therefore, sought to assess the solar energy financing models that could increase household accessibility to solar energy. It also evaluates the socio-economic and environmental indices that could sustainably lead to an energy transition in the Ngaciuma-Kinyaritha sub-catchment.

1.2 Statement of Problem

While there is little doubt about the size and growth rate of the Kenyan solar energy market, there is still an ongoing debate about how to interpret the significance of the current financing models for SHS, with emphasis on economic and environmental relevance. Kenya's population as at 2011 was estimated to be 41 million inhabitants with projected growth rate of 2.7% per annum. This growth is accompanied by increased household demand for energy leading to an energy deficit of about 3,000 megawatts notwithstanding its current production of 1,100 megawatts. Owing to this deficit, the current policy document of Kenya, the Vision 2030, highlights a search for alternative means of providing sustainable energy to meet both its rural and urban development aspirations. Despite the high rate of urbanization in the country, the rural sector still retains over 70% of the total population, who depend mainly on woodfuels and paraffin to meet their basic energy needs. These sources are known to reduce rate of carbon sequestration and increase Green House Gas (GHG) emission. The advent of solar technology seems to be changing this energy use pattern by displacing household dependence on paraffin, with about 20% growth rate in the number of SHS installations each year. In Ngaciuma-Kinyaritha sub-catchment, Kenya, about 6.6 percent of households have transited to the use of SHS as at 2009. Despite the relevance of these figures, it is contended that the economic and environmental benefits of SHS are relatively insignificant owing to the predominance of low capacity systems. Studies show that the initial capital requirement remains the leading obstacle to access and use of higher capacity SHS. This resonates with the argument that the current models of solar financing are not optimizing the potential of solar energy in rural communities including Ngaciuma-Kinyaritha sub-catchment. It was therefore deemed useful to evaluate the economic and environmental significance of various financing models for SHS and to analyse different scenarios in order to determine sustainability financing models that together overcome the cost challenge of solar energy use in Ngaciuma-Kinyaritha sub-catchment.

1.3 Justification and Significance of the Study

Environmental sustainability has become a top issue in development agenda, at the international, regional and local level. With regard to energy use, the focus has shifted from fossil fuels to Renewable Energy Technology (RET) which generates less or zero environmental externalities. Apart from its stand alone capability, solar energy is known to have the least environmental effect and has widely been adopted by rural households in developing countries including Kenya. The key inhibiting factor to solar energy, the cost challenge, is therefore worth scrutiny in order to define different trajectories for promoting solar energy and sustainable development. This study provides insight into the financing spectrum of solar energy in Ngaciuma-Kinyaritha sub-catchment, Kenya. It recommends action-oriented interventions for solar energy financiers, policy makers and all stakeholders who play a role in the energy or environment framework. It also include the environmental consciousness dimension of energy use in rural households and recommends specific strategies for monitoring and enhancing environmental consciousness in energy use. The findings from the study provide alternative futures of household solar energy financing through scenario analysis which is very useful in policy formulation and decision making.

1.4 Research Questions

i. How are household energy needs met in Ngaciuma – Kinyaritha sub catchment?

ii. Which financing models are the most effective for solar energy dissemination in Ngaciuma – Kinyaritha sub catchment?

iii. What is the level of environmental consciousness of energy use by households in Ngaciuma – Kinyaritha sub catchment?

1.5 Research Hypothesis

This study was guided by the following hypothesis which are stated in the null form:

1) The cost of solar energy does not limit considerably accessibility by households in Ngaciuma-Kinyaritha sub-catchment.
2) Increased solar energy access by households has no significant effect on environmental degradation in Ngaciuma-Kinyaritha sub-catchment.

1.6 Research Objectives

1.6.1 General objective

The central objective of the study was to evaluate the economic and environmental significance of different solar energy financing models and to analyse different scenarios in order to provide sustainable financing models for solar energy in Ngaciuma-Kinyaritha sub-catchment, Kenya.

1.6.2 Specific objectives

i. To characterize the different sources of energy used by households in Ngaciuma – Kinyaritha sub-catchment.

ii. To assess the cost benefit ratios of different models of solar financing in Ngaciuma – Kinyaritha sub-catchment.

iii. To assess the environmental consciousness of energy use by households in Ngaciuma – Kinyaritha sub-catchment.

1.7 Scope and limitations of the study

The contextually scoping of this study is defined by the IWM concept focusing the beam on the environmental, economic and social aspects of energy use in Ngaciuma-Kinyaritha sub-catchment, Kenya. Regarding environmental and social aspects, the key objective was to determine the household environmental consciousness which according to Reddy (2008), is regarded as the impetus of sustainable energy transition. The study however lends priority to the cost of solar energy and the various financing models making use of scenario analysis to determine alternative trajectories towards sustainable energy use in the sub-catchment. Despite its attempt to investigate and deliver the stated outputs, this study was constrained by resource inadequacies including time and financial limitations.

1.8 Operational Definition of Terms and Concepts

The key terms that should be defined in the context of this study include: Model, Environmental Consciousness and Knowledge Level. In order to give the study a contextual scope, the following definitions are adopted in this study.

1.8.1 Financing Model

The adopted definition for a model in this study is synthesized as a representation of a system that allows for investigation of the properties of the system and, in some cases, prediction of future outcomes (Kuhne, 2005). In the context of this study, financing model is used to depict any representation of financing event in the real world which contains the following key features:

a. Necessarily incomplete; does not include every aspect of the real world.
b. May be changed or manipulated with relative ease. This characteristic makes it easier to study a model than the real world event.

1.8.2 Environmental consciousness

The term environmental consciousness is used in reference to the *"desire to protect flora and fauna, a willingness to scrutinize the consequences of economic activity and a willingness to combine long-term with short-term planning"* (Hampel and Holdsworth, 1996, pg 6). In the context of this study environmental consciousness is used to refer to variables such as knowledge, awareness, alertness and willingness to make environmentally proactive choices. These variables of environmental consciousness are estimated with the help of a graded scale (Likert scale) and used as indicators of household environmental consciousness.

1.8.3 Scenario Analysis

Scenario analysis has been used by public sector agencies, private sector organizations and non-Governmental organizations to manage risks predict future outcomes and develop strategic plans in the midst of uncertainties (Maack, 2001). Scenario analysis as a decision making tool in this study is used to build alternative financing environments and determine their respective rates of success using PESTELI. It categorizes the observed macro environmental factors that enhancing or inhibit access and use of PV solar home systems into Political, economic, socio-cultural, technological, environmental, legal and institutional factors.

CHAPTER TWO

LITERATURE REVIEW

2.1 Introduction

This chapter presents a review of relevant literature based on the following thematic areas: the role of alternative energy in sustainable development, Energy policy framework for sustainable energy transition, financing models of Solar Home Systems (SHS), the link between energy use and environmental conservation and finally a conceptual framework for the study. At the end of the review, knowledge gaps have been identified and suggestions of how the present study did contribute to fulfilling them are outlined.

2.2 Role of Alternative Energy in Sustainable Development

According to IEA (2002), there is no single universal definition for alternative energy owing to a variety of energy choices and differing goals of their advocates, defining some energy types as alternative is highly controversial (Nuclear energy). For instance, MacKay (2007) remarks that "alternative is an umbrella term that refers to any source of usable energy intended to replace fuel sources without the undesired consequences of the replaced fuels". IEA (2002), maintains that energy obtained from sources that are unlimited, rapidly replenished or naturally renewable are termed as alternative sources of energy. From the above, this study notes that alternative energy is environmentally-friendly source of energy that easily replenishes itself to sustain the supply of the energy needs of consumers.

According to the World Bank Group (2009: p.2) global energy-related carbon dioxide emissions will increase by about 50 per cent between 2004 and 2030 unless major policy reforms and technologies are introduced to utilise alternative energy. Thus, renewable energy technologies hold the answer to the incidence of global warming (IEA cited in UN-Energy, 2008, World Bank Group, 2009). United States Department of Energy (2001) espouses that the main forms of alternative energy are the wind power, hydropower, solar energy, bio-fuel and geothermal energy.

According to the Alternative Energy Institute (AEI) (n.d.), solar energy had been neglected due to the availability of fossil fuels that were more affordable and available. However, with the growing concern for the environment coupled with the increasing cost of fossil fuel exploitation, attention has been given to solar energy (Hilling, 2011). Direct solar energy can broadly be categorized into solar photovoltaic (PV) technologies, which convert the suns energy into electrical energy; and solar thermal technologies, which use the suns energy directly for heating, cooking and drying. Photovoltaic solar power is one of the most promising alternative energy sources in the world (Jacobson, 2007). The 89,000 TW of sunlight reaching the Earth's surface is plentiful – almost 6,000 times more than the 15 TW equivalent of average power consumed by humans. Solar power is pollution-free during use. Assuming that our rate of usage in 2005 remains constant, estimated reserves are accurate, and no new unplanned reserves are found, we will run out of conventional oil in 2045, and coal in 2159 (GENI, 2008).

In 2007 grid-connected photovoltaic electricity was the fastest growing energy source, with installations of all photovoltaics increasing by 83% in 2009 to bring the total installed capacity to 15 GW (AEI, 2010). Nearly half of the increase was in Germany, which is now the world's largest consumer of photovoltaic electricity followed by Japan. AEI (n.d.) argues that the efficiency of PV has improved considerably over the years through research and this has reduced the cost of installation. Ishengoma (2002), however, contends that the two main obstacles against using solar energy are the high initial capital costs and the very low PV cell conversion efficiency.

Despite the multi-functionality of the solar energy that the above paragraphs illustrate, it is tempting to buy into the well established fact that the complex patterns of use and dynamics of fuelwood in the rural household cannot be dealt with by 'simplistic' linear models (Mahiri, 2003). This in a sense suggests that simply making solar energy widely available and accessible is not a panacea to woodfuel consumption. Inferentially, the issue of energy demand may require a more comprehensive rather than a tentative intervention. This study thus sought to provide an answer to the question, how satisfactory is the household environmental consciousness of energy users in the midst of these complexities? Reviewing of the practical inadequacies of environmental economics, Reddy (2008) emphasized the real essence of environmental consciousness as an influential catalyst of green consumerism. From IWM view point, the new dimension that this study brings with regard to assessing household energy use is estimating Household environmental consciousness of energy use and the role it plays in a sustainable energy transition.

2.3 Policy Framework for Sustainable Energy Transition

The relatively progressive financial sector that has incorporated photovoltaic equipment into its consumer goods portfolio coupled with the removal of import and foreign exchange controls constitute the major impetus of the solar industry in Kenya (Wamukonya *et al.*, 2002; Green, 2002; Hankins, 2010). However, poor standardization pervades the solar energy industry in Kenya. According to Jacobson (2001), the solar industry suffers from erratic equipment and installation standards - "Kenyan manufacturers make more than 90 percent of the batteries used in local solar home systems, 30–50 percent of the lamps, and perhaps 10 percent of the charge regulators". Due to the liberalization of the energy market in Kenya, it is difficult to regulate the manufacturing or distribution process and thus, the quality of solar home systems is not fully assured in the Kenyan market. It is worth noting that the financing modalities have the potential to influence the quality of products on the market.

The Kenya Vision 2030, the current policy document of Kenya, indicates that energy transition is primal to the realization of the socio-economic pillars within the development framework of the Vision. It stipulates that the government is committed to continued institutional reforms in the energy sector and that "new sources of energy" will be found through exploitation of both renewable and non renewable forms of energy (Ministry of Energy, 2011). The vision recognizes the fact that energy ties together the comprehensive progress of all the remaining pillars. In line with this, the ministry of energy is taking steps to integrate renewable energy use into the energy mix. One of such visible steps is the Scaling-Up Renewable Energy Program in Low Income Countries (SREP) of which Kenya is among the six pilot countries to benefit (Ministry of Energy, 2011). Hitherto, most rural catchments including Ngaciuma Kinyaritha still depend on Paraffin and woodfuels to meet their daily energy needs(Table 2.1).

Table 2.1 Energy use in Ngaciuma Kinyaritha

	Household distribution by main cooking fuel (%)	
	Firewood	86
	Paraffin	4.5
	Electricity	0.2
	LPG	0.8
	Charcoal	6.8
	Biomass residue	0.1
	Others	1.4
	Household distribution by main lightening fuel	
	Firewood	2.0
	Grass	1.1
	Paraffin	76.8
	Electricity	12.6
	Solar	6.6
	Dry cell torch	0.6
	Candles	0.2
	Household distribution by cooking appliance type	
	Traditional stone fire	62.4
	improved traditional stone fire	21.5
	Ordinary jiko	4.0
	Improved jiko	5.1
	Kerosene stove	4.2
	Gas cooker	0.8
	Electric cooker	0.2
	Others	1.6

Source: Imenti North District Development Plan, 2008

To attract private sector capital in bulk solar electricity generation, the energy policy of Kenya indicates that a Feed-in-Tariff of US Cents 20.0 (17shillings) per Kilowatt-hour will be made available for a bulk solar energy producer in the first 20 years. This is intended to be used to supply the off-grid stations, include Lamu, Lodwar, Mandera, Marsabit,Wajir, Merti, Habasweni, Elwak, and Baragoi and others (Ministry of Energy, 2011). According to Gueye *et al.* (2004), the simultaneous support from the World Bank and multilateral funding organizations is seen as an essential replacement of subsidized rural electrification with unsubsidized or less subsidized market based PV systems.

UNEP (2006) observed that social and environmental issues in energy use are only beginning to take root in Kenya. Given the growing recognition, the Integrated Assessment and Planning (IAP) initiative in Kenya generated three integrated scenarios (*Business As Usual, Implementation* and *Win-Win scenarios*) as an analysis tool to assess the energy planning process in the current energy development framework. The process identified the following as the major challenges to the energy sector and as such, solar energy provision: inadequate human and financial capacity, paucity of data, weakness in the budgetary process, lack of institutionalized planning process, ineffective inter-ministerial coordination and many more. All these inadequacies are paralleled through the solar energy industry as well and therefore warrant attention of research in order to provide comprehensive insight for sustainable energy transition.

2.4 Financing Models of Solar Energy in Kenya

Among other financing approaches for solar energy, market-based rural electrification is increasingly becoming a household approach in developing countries (Jacobson, 2006, UNEP, 2010). To this effect, Kenya may have taken the lead in east Africa in. However, solar electricity still faces some major limitations gluing large populations in the Ngaciuma Kinyaritha subcatchment to the use of non sustainable and non environmentally friendly energy sources (Karekezi and Kithyoma, 2003; Mahiri, 2003),. This study seeks to establish among others, how 'alternative' is solar energy for household.

Owing to the *Electric Power Act -1997which was recently replaced by the Electricity Act-2007*, rural grid connection programs have been deprioritized since it is regarded as not being cost-effective (Wamukonya *et al.*, 2002). The same Act which liberalized the energy sector also led to poor accessibility of LPG in most rural areas leaving rural Kenya handicapped in energy supply (Jacobson, 2006). This, though indirectly, provided the needed leverage for growth in demand for solar energy in Kenya.

Though the solar industry still suffers from erratic equipment and installation standards, the major challenge for PV dissemination in Kenya, remains the high initial cost of PV products (Jacobson, 2006). Typically, the cost components of a solar system involves a solar panel which accounts for 60% of total cost; Mounts and wiring which accounts for 15% of total cost; an inverter which accounts for 10% of total cost and installation workmanship which accounts for 15% of total cost. Installation of SHS in Kenya typically costs between US$500 and US$1200 depending upon size, components and taxes rebates. A higher capacity unit may range from 3kw to 10kw usually average between US$12000 and US$60, 000 (Cory and Coughlin, 2009; Gueye *et al.*, 2004). Guaye *et al.* (2004) argues that owing to the fact that rural PV buyers are engaged in agricultural production and do not have sufficient cash in hand, most users go after the cheapest PV equipment available in the market.

Both Owusu (2010); Gueye *et al.* (2004) outlined the different models of financing solar energy. These include: cash sales, dealer credit, end-user credit, hire-purchase and Solar Developer. Among these models, Gueye *et al.* (2004) observes that the end user credits is the most efficient with respect to meeting the needs of the rural poor. It involves receiving credit from a "third party" credit provider, such as micro-finance organizations or NGOs. This model also permits users to maintain ownership of SHS and be responsible for maintenance and repair.

Aside from the above, Cory and Coughlin (2009) highlight some "new financial models" which included third party ownership model, community joint financing as well as monetizing environmental values. Among these three, third-party ownership models is seen as the most efficient model where a Solar developer takes advantage of government tax incentives to provide solar energy from a local production station and supply it by grid to homes and institutions. This model eliminates the up-front costs as well as operations and maintenance responsibilities but it has not been explored in the growing solar industry of Kenya. The scepticism that these models are not feasible alternatives in the Kenyan solar energy market is justifiable within the boundaries of the paucity and general inadequate information. Thus, this study seeks to provide information for evidence-based decision making with regard viability of alternative solar energy financing models to foster the drift from woodfuel and paraffin to solar energy. Table 2.2 is an illustration of the framework that this study adopts as a guide for remodelling solar energy financing in Ngaciuma Kinyaritha sub-catcment based on income classes and household energy requirements.

Table 2.2 Framework for Remodelling Energy user-financing

Off Grid Homes and SME Solar Systems (less than 1 kW) -run small home appliances, street lights, water pumps etc	**Low income consumers (individual)**	**Middle income consumers (individual)**	**High income consumers (individual)**	**Local Institutions**
	Cash sales, dealer credits, third party credits, Sola developer	Cash sales, dealer credits, third party credits, Sola developer	Cash sales, dealer credits, third party credits, Sola developer	Cash sales, dealer credits, third party credits, Sola developer
Off Grid Homes and Ranches (1kW to 5kW)	Low income consumers (Group-based)	middle income consumers (Group-based)	High income consumers (individuals)	Local Institutions
For Institutional, Industrial and production demands	Cash sales, dealer credits, third party credits, Sola developer	Cash sales, dealer credits, third party credits, Sola developer	Cash sales, dealer credits, third party credits, Sola developer	Cash sales, dealer credits, third party credits, Sola developer

Source: Adopted and modified from Cory and Coughlin (2009)

2.5 The Relationship between Solar Energy and Environmental Conservation

Different studies (Karekezi & Kithyoma, 2003, AEI 2010) have indicated that emission from biomass energy sources, including indoor air pollution from unvented biofuel cooking stoves, is one major contributor to respiratory illnesses in highland areas of sub-Saharan Africa. Also, as populations grow and oil prices increase more and more people in the developing world will continue to cut down trees for firewood and charcoal reducing total forest resources as well as the value of ecosystems (Schulte-Bisping H. *et al.,* 1999; Sohngen, 2008).

In developing countries where majority of the population is employed in the agricultural sector, deforestation means more than the loss of forest resources. Deforestation means less arable land, low soil productivity influenced by soil erosion and increased predisposition to the vulnerabilities of food insecurity (Fischer *et al.,* 2007). Vegetation cover has been shown to be a major determining factor in the control of erosion by influencing the soil hydrology through interception, increased infiltration, and evapotranspiration (Table 2.3). Obando (2005) establishes that erosion rate (*E*) is a function of overland flow, surface gradient or slope, and vegetation cover by using the mathematical model illustrated below.

Model for Estimating Soil Erosion

$E = Kq^{m}\ S^{n}\ e^{-bVc}$, where;

K (dimensionless) is a soil parameter, which describes the erodibility of the soil.

q is the discharge (m^3/s)

S is tanβ where β is the slope in degrees

Vc is the percent vegetation cover

E is the erosion (mm/m)

b m and *n* are dimensionless parameters, *b* relates to the reduction in erosion due to vegetation cover.

Source: Obando (2005); Modeling Soil Erosion and Vegetation Change

However, measured against individual economic benefits, farmers as well as wood fuel producer and consumer find no incentive for biodiversity preservation and emissions lowering. It is in line with this that Reddy (2008) and Tantawi *et al.* (2009) argue that a mere ethical standing would not be sufficient incentive an energy transition. Again, this is a question of environmental consciousness; the knowledge, awareness and willingness to make environmentally-proactive choices in energy use (Sánchez and Lafuente, 2010). With regard to evaluating the forms of energy used in the sub-catchment, the study brings to light the environmental consciousness dimension of energy use and the household propensity to transit to using solar energy.

2.6 Conceptual Framework

This study is grounded on a set of concepts that have a bearing on key variables; cost benefit analysis of selected SHS, willingness and ability to pay for SHS, potential change in solar energy demand and environmental effect of solar energy use. It begins with the concept of environmental externalities of production and consumption and the need to minimize these externalities by transiting to renewable forms of energy which have little environmental effect (Pigou, 1939). However, as Jacobson (2006) argues along with Karekezi and Kithyoma (2003) both the environmental and economic significance of solar energy is perceived to be minimal due to predominance of low capacities and that in the absence of large subsidies, solar energy is primarily for the few rural elite. With regard to this, many studies (Gueye *et al.*, 2004; Cory and Coughlin, 2009; Owusu 2010) reviewed in the above literature conclude that the point of intervention is hinged on the financing models for solar energy. It also show, as in Figure 2.1, that the current models of financing for solar energy which is weighed down by initial capital upfront and low capacity systems helplessly cut off access by the extremely poor and generates less economic and environmental payback.

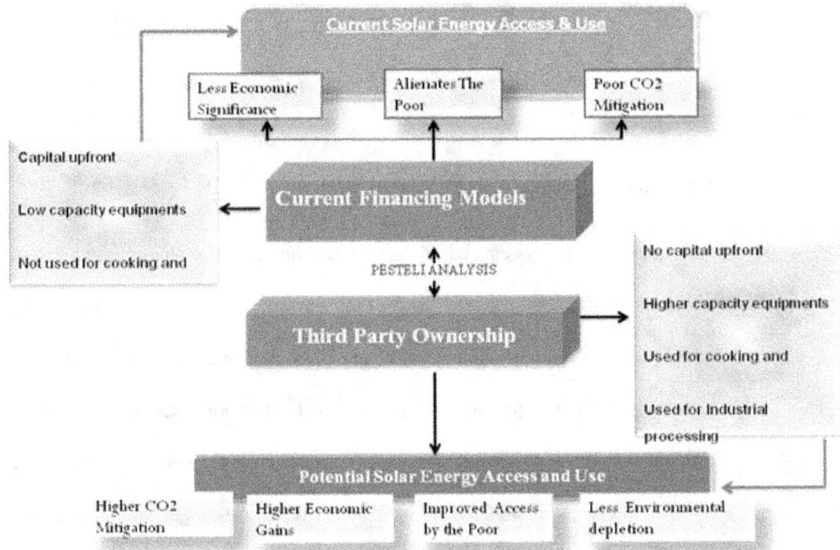

Figure 2.1 Conceptual Framework of the Study: Alternative models of solar energy
Sources: Synthesised from Cory and Coughlin, (2009), Jacobson, 2006 and Gueye et al., 2004.

In the bid to evaluate these financing models, the study relies more on the concepts propounded Cory and Coughlin (2009), as well as Gueye *et al.* (2004), which brings to focus the traditional and non-traditional financing models. As illustrated in Figure 2.1, the study explored the current financing models as well as some potential financing models with the aid of scenario analysis. The scenarios tested this study portends to reign in the poor accessibility to solar energy and to increase the economic and environmental payback by leveraging on the absence of initial capital upfront and a considerable willingness to pay.

2.7 Chapter Summary

Sustainable energy systems based on solar technology offer an opportunity to protect the environment and create economic growth. However, the success of Renewable Energy Technologies- RETs (including solar) in Africa has been limited by a combination of factors which include (a) high initial capital costs, (b) poor baseline information, (c) weak maintenance service and (d) poor infrastructure. This study addressed the primal challenge which is meeting the initial cost of solar energy through a cost benefit analysis for different models of solar energy financing based on the current cost of available solar home systems or PV systems. It would provide among others the benefit/cost ratios of selected systems, the payback period, the life payback and the best models of financing for each selected system.

CHAPTER THREE

MATERIALS AND METHODS

3. 1 Introduction

This section details the scientific pillars of the study with emphasis on the theoretical background, research design, sample size and the sampling techniques used. A case study approach is used to integrate both empirical and constructive perspectives of this research laying emphasis on the existing knowledge and inherent relationships in a community that together influence an energy transformation process.

3.2 Study Area

Ngaciuma-Kinyaritha is a small catchment of 167 km^2 in Imenti North District in Eastern Province of Kenya. It has a population of about 36,000 people, representing a density of approximately 360 persons/km^2 as at 2009. The catchment is endowed with relatively substantial natural resources including the Imenti forest, well drained soil, rainfall (about 1100 mm in the lower zone to 1600 mm in the upper zone) and sunshine (DAAD, 2007; Imenti North District Development Plan, 2008). The Ngaciuma-Kinyaritha Sub-catchment forms part of a larger pilot study on watershed management undertaken by the Integrated Watershed Management network. According to the Imenti North District Development Plan, 2008, this sub-catchment is noted for growing popularity PV solar energy use. Thus is its selection for this study.

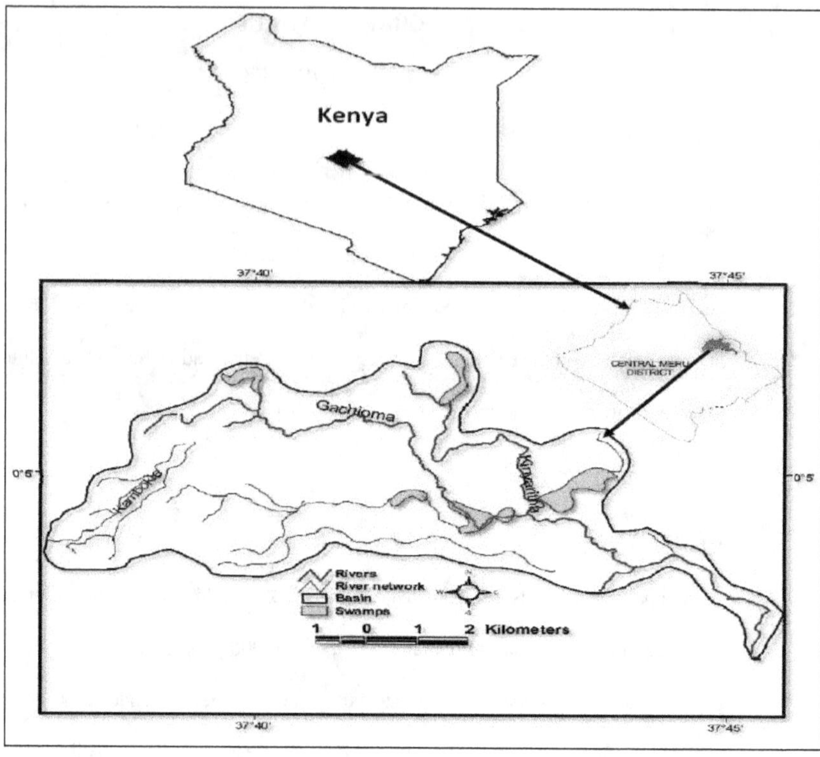

Figure 3.1: Map of Ngaciuma Kinyaritha Sub-catchment
Source: Survey Map of Kenya, Sheet No. NA-37-14, (1996)

It is estimated that about 86 percent of households depend on wood for cooking and 76% on paraffin for lightening (Imenti-North District Development Plan, 2008). The remaining population or 24% and 34% rely on charcoal, electricity, paraffin, LPG and others (Appendix 1.0). Also, the energy choice shows that demand for cooking fuels vary depending on whether the household is located in rural or urban areas and is also driven by certain key factors including household size, price of charcoal, price of fuel wood, education level, and both formal and informal employment (Imenti-North District Development Plan, 2008). Table 3.1 illustrates the climatic conditions in the sub-catchment for the three zones.

Table 3.1 Geographic Features of Ngaciuma-Kinyaritha Sub-catchment

Zone	**Area (km²)**	**Rainfall amount (mm)**	**Evaporation (mm)**	**Max temp. (°C)**	**Max temp. (°C)**
Upper	13	1,300	1,300	22	11
Middle	27	1,000	1,700	26.8	15
Lower	109	700	1,800	29.6	17.3

Source: DAAD (2007)

3.3 Selection and Training of Research Assistants

Three research assistants were selected based on good English and Kiswahili speaking, and a basic understanding of field research to assist in the translation, field recording during Focused Group Discussions (FGDs) and general community organizing and guidance. The three assistants were given on-field training using the survey instruments designed for the study. This was necessary in order to anticipate any challenges during the field enumeration and to facilitate an efficient field data collection. During the field enumeration, the research assistants were engaged in organizing a cognizance survey, administering some questionnaires (with supervision by researcher), translating and recoding responses during the FGDs.

3.4 Key Variables and Sample Units

The key variables that the study focused on include the (1) cost benefit ratios of selected SHS, (2) the ability and willingness of households to pay for solar energy under the different models of financing, (3) potential increase in Solar energy demand under the different financing models and (4) potential consumer effectiveness of solar energy on environmental conservation. Figure 3.2 illustrates these key variables.

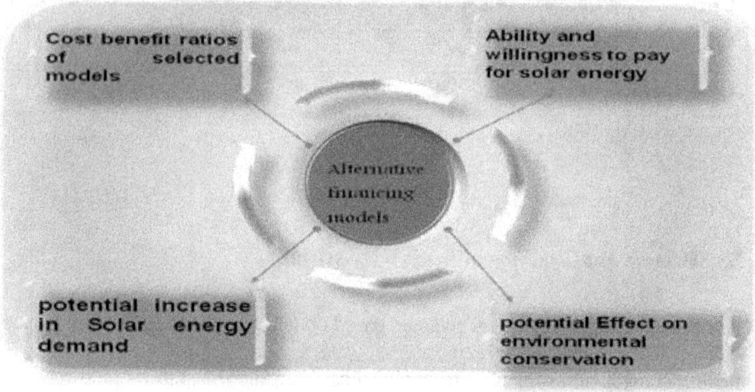

Figure 3.2 Key Variables of the Study
Sources: Developed from Gueye et al., 2004; Jacobson, 2004; Mahiri, 2003

3.5 Research Design

3.5.1 Sampling procedure

Kumekpor (2002) states that a multi-stage sampling methodology is relevant for studies that do not have a more elaborate sample frame (Kumekpor (2002); Flyvbjerg (2006).

Based on this strength, the proposal stratified the study area into the upper zone or cluster, the middle zone or cluster and the lower zone or cluster. Figure 3.3 illustrates the sampling of procedure of the study. The study adopts Yamane's, (1967) formula for population sampling to derive its sample size (Yamane, 1967:886 cited in Glenn, 1992). This is given as follows:

$$n = \frac{N}{1+N(e)^2} \qquad \text{(Equation 1)}$$

Where n= Sample size, N= Population size, e= Margin of error

The zoning was premised on the fact that, inasmuch as the activities of each community affects the entire catchment, the relative effect on communities upstream is different from that of communities downstream (DAAD, 2007; Förch and Ngonzo, 2009). Based on equation one (1), the sample population of each cluster was determined, totalling 100 households (Figure 3.3). The 100 samples households were distributed proportionately among 9 randomly selected communities for the upper, middle and lower zones. This technique ensured that every sample element has equal opportunities of being selected. In each of the three zones, 10 households were selected for an FGD to provide insight for analysing scenarios generated.

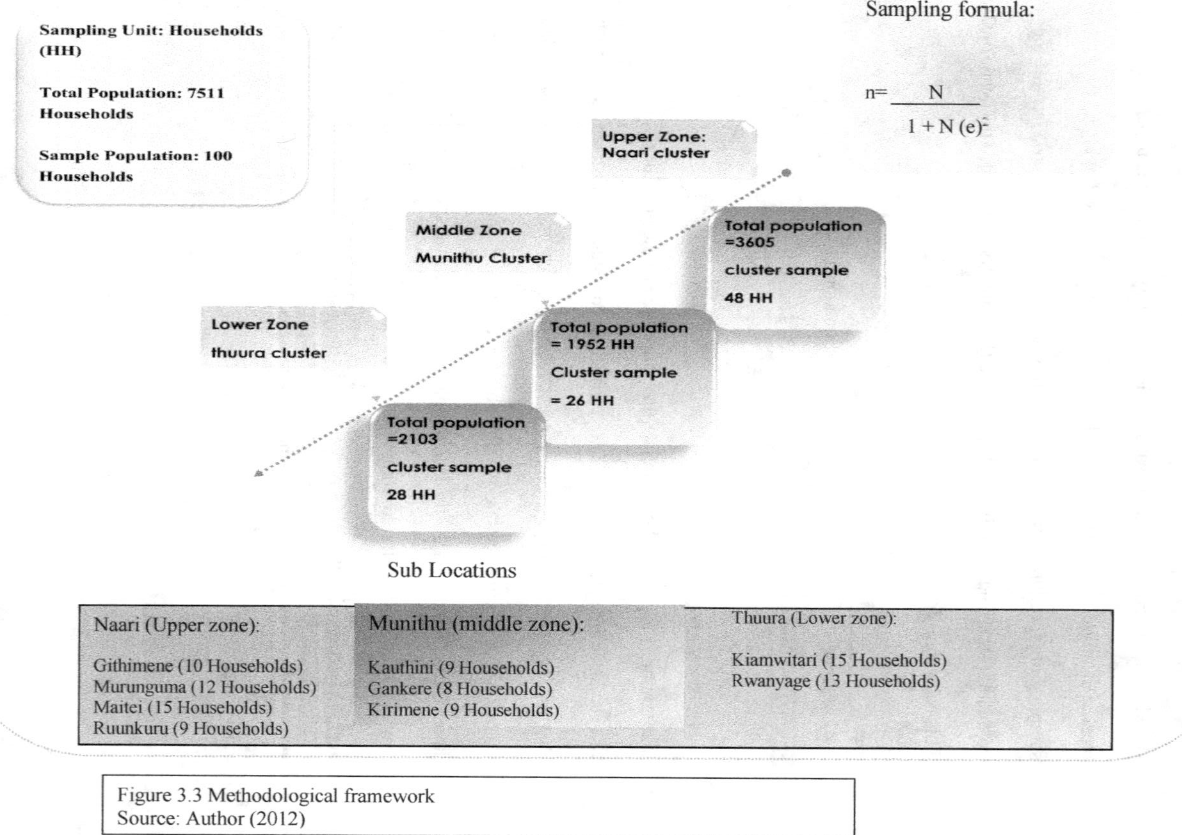

Figure 3.3 Methodological framework
Source: Author (2012)

3.5.2 Sources of data and methods of data collection

Since the aim of this study was to assess the economic and environmental relevance of solar energy in Ngaciuma Kinyaritha sub-catchment to foster a sustainable energy transition, the study also gathered second-hand data from published and unpublished materials to provide a supporting theoretical framework for guided study. The secondary sources included journals, scientific reports, dissertations, websites, and other relevant sources of material information relevant to solar energy financing.

Primary data was collected using questionnaires (Appendix 2.0), in-depth interviews (Appendix 3.0) and focused group discussions; composed of groups of solar energy user-households, wood and paraffin dependent households and a Water Resource User Association (WRUA) in selected communities. Based on the reviewed financing models in this study, three alternative scenarios were generated and field tested in the sub-catchment. These scenarios include third party credits, joint community procurement and third party ownership or solar developer models. To ensure a consistent data collection and a systemised data analysis, the nested knowledge relationships approach as a guiding framework for both stages was adopted (Figure 3.4). The nested knowledge relationships approach refers to a process used for modelling knowledge units within communities required in moving towards sustainable environmental management as illustrated in figure 3.3.

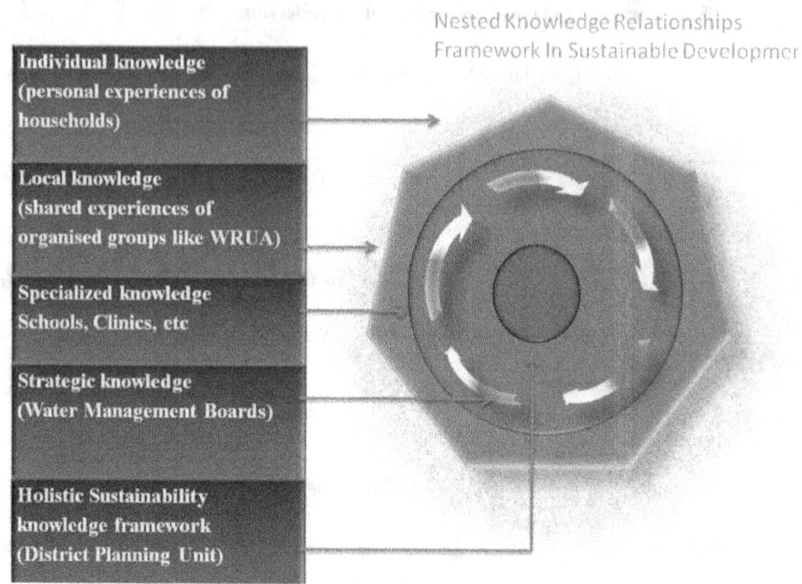

Figure 3.4 Nested Knowledge Relationship Approach to Sustainable Watershed Management

Sources: Norman (2002); Keen et al. (2005)

It begins with individual experiences and local shared experience and then integrates knowledge from specialized institutions like finance, health, education as well as knowledge from local governance agencies (Figure 3.4). These together operate within a holistic sustainable development framework of the country. Therefore data triangulation technique was also used to verify the institutional, household and local knowledge obtained. Owing to the application of different data collection tools, triangulation served as a means of harmonising the various results into a single investigation.

3.5.3 Ethical considerations of the study

In recognition of appropriate research principles, the study conducted under anticipate ethical standards in order not to go against the rights of its participants. First, the study obtained permission from the Meru Municipality and the WRUA, which is the main local body in charge of environmental management. It also recruited one local community member to guide the interactions with the participants in order to avoid running into unanticipated ethical circumstances. The ethical bearing of this study was especially useful in the FGD sections and it ensured active engagement of participants and respondents.

3.5.4 Confidentiality of the study

The entire study was conducted under the assumption that its findings would be presented anonymity. For this purpose, each survey instrument introduced the purpose of the study to the respondent seeking the permission of respondents and indicating that the respondent is not obliged to address every item in the instrument. Also, each case in the survey was assigned a unique identification so respondents had the free will to provide any personal details. The study also sought for the permission of participant before including any personal details like names and images of any kind. Interviewers assured the participants that these details were only meant to support this academic study and not for any other purpose. It is therefore crucial that the output of the study represent only the observations that were obtained under this protocol.

3.5.5 Pilot study

As part of the strategies of the study to ensure logistical efficiency as well as careful use of resources the study undertook a pilot study in which it tested the survey instruments that had been designed. This pilot study also gave the both the researcher and research assistance the geographic and socio-cultural cognisance which was useful in interacting with the communities. This exercise also served as a hands-on examination exact framing of questions, whether they fetched the right responses, whether the questions were adequate or whether it contained any irrelevant items. In the end, the analysis of the pilot study revealed missing items in the household questionnaires that were field in before the study commenced.

3.6 Data Analysis Procedure

The major statistical package that was used in this study is the Statistical Package for Social Sciences (SPSS 17.0). This was supported with Microsoft excel and other web-based softwares such as Qlickview. SPSS would be used to test the hypothesis and to generate descriptive statistics including frequencies, means, percentages, as well as cross tabulations. Descriptive statistical analysis in this study essentially dealt with simple measurements such as frequencies, percentages, means, minimum and maximum values, range and standard deviations. These measurements provided scientific understanding of the various forms of energy used to meet household energy needs. It also provided insight into the various financing models for solar energy in the study area. The study also employed Chi-square analysis due to the qualitative aspects of the study (Equation.2). This is calculated as using the following formula (Wilson and Hilferty, 1931):

$$X^2 = \sum_{i=1}^{n} \frac{(O_i - E_i)^2}{E_i}$$ (Equation.2)

Where;

X^2= Pearson's cumulative test statistic, which asymptotically approaches a χ^2 distribution.

O_i= an observed frequency; 4

E_i= an expected (theoretical) frequency, asserted by the null hypothesis;

n= the number of cells in the table.

3.6.1 Computing carbon mitigation due to energy transformation

Different accounting methods for carbon credits for CDM forestry projects have been summarized and discussed by IPCC (2000). The ton-year accounting method is one of the most commonly used and it assumes one credit would be awarded to a project that stores 1 ton of CO^2 during a period of 100 years or Te years. Under the 'ton-year' method, the amount of credits issued to a project during a given period of time is calculated as follows(Equation.3) (adopted by IPCC, 2000):

$$CER_{ton-yr} = \frac{\sum_{t=x}^{t=x+i} (CO_{2project} - CO_{2baseline})}{Te}$$ (Equation 3)

Where;

Te is the 'equivalence time',

X is the beginning of a crediting period, and

i is its duration in years.

Owing to the low emission levels from rural households, this study used the emission factors in Table 3.2 to calculate the carbon mitigation benefit attributable to households for due to solar energy use. In Kenya the cost of carbon mitigation from solar electrification ranges from US$110 to US$140 (Jacobson, 2007). This study used US$140 or 8200Ksh as the benefit attributable to households for carbon mitigation due to solar energy.

Table 3.2 Emission Factors by Type of Fuel

	Fuels (g per kg of dried fuel)				
Emitted Substances	LPG	Kerosene	Charcoal use	Firewood	Residues
CO_2	3085	2943	2706	1580	1426
CH4	0.05	1 .1	7 .9	2 .8	-
N2	O 0.15	0.1	0.2	0.07	-
PM	0.51	0.7	2 .4	0.9	-
CO	14 .9	62	135	7 0.9	67 .4

Source: Energy for Sustainable Development Africa, ESDA (2003)

3.6.2 Cost benefit Analysis solar energy financing

This study made use of cost benefit analysis to determine the cost-effectiveness. It compared the relative costs and effects of different solar energy investments at the household level assigning monetary value to the measure of effect (Vishnudas *et al.* 2005). The analytical procedure was based on the principle of the value of money which involves discounting for selected PV solar modules investment over an assumed life span of 25 years.

3.6.3 Calculating the Payback Period of solar PV systems

The time value for money is a central financing concept that allows the valuation of likely streams of income in the future in such a way that the annual incomes are discounted and then added together, thus providing a lump-sum present value of the entire income stream. With the aid of discounting method, the study computed the payback period and the life payback of an investment in solar energy as a basis for comparing the cost effectiveness of different investment scenarios (Equation.4). The discounting formula for was use to calculate the present values (Weitzman, 1998 cited in Gollier, 2009:

$$PV = \frac{FV}{(1+i)^n} \qquad \text{(Equation 4)}$$

Where;

PV is the value at time=0

FV is the value at time=n

i is the discount rate, or the interest rate at which the amount will be compounded each period

n is the number of periods (not necessarily an integer)

The benefit-cost ratio (BCR) is calculated as the NPV of benefits divided by the NPV of costs (Weitzman, 1998 cited in Gollier, 2009):

$$BCR = \frac{\sum_{t=1}^{T} \frac{B_t}{(1+r)^t}}{\sum_{t=1}^{T} \frac{C_t}{(1+r)^t}} \qquad \text{(Equation 5)}$$

Where B_t is the benefit in time t and C_t is the cost in time t. If the BCR exceeds one, then the project might be a good candidate for acceptance.

3.6.4 Scenario Analysis with PESTELI

The external environment of a development project, organisation or community can be assessed by breaking it down into Political, Economic, Social, Technological, Environmental, Legal and Industry levels; termed PESTELI analysis. PESTELI analysis is a useful tool for understanding the broad framework or environment in which an individual or group is operating- the internal and external enhancing and inhibiting factors. This facilitates the process of making use of the available opportunities to minimize the threats in that environment. Table 3.3 provides a guide for conducting PESTELI analysis.

Table 3.3 Scenario Analysis Guide

Insert Subject for PESTELI analysis:	
Political	**Economic**
Government type and stability Regulation and de-regulation trends Tax policy, and trade and tariff controls	Current and projected economic growth, inflation and interest rates Likely impact of technological or other change on the economy Likely changes in the economic environment
Socio-cultural	**Technological**
Population growth rate and age profile Population health, education and social mobility, and attitudes to these	Impact of emerging technologies Impact of Internet, reduction in communications costs and increased remote working
Ecological factors	**Legislative requirements**
Air quality, transportation, parking, pollution discharge, water quality, waste management.	Primary and secondary legislation in relation to Health Bills

Institutional analysis
Demand, liaison and selection for services, products and/or component parts on the basis of price, quality.

Source: Maack (2001)

Scenario analysis has been used by public and private sector organizations to predict future outcomes and develop strategic plans in the midst of uncertainties (Maack, 2001). Scenario analysis is used in this study to build alternative financing environments and determines their respective rates of success using PESTELI.

Table 3.4 Summary of Research Methodology

Study Objectives	Key variables	Method of Analysis
To evaluate the different sources of energy used by households in Ngaciuma-Kinyaritha sub catchment.	Source of energy used for lighting, Source of energy for lighting Ability to pay	Descriptive statistics; frequencies and cross tabulations, Pearson's Chi-Square Test
To assess the environmental consciousness of energy use by households in Ngaciuma-Kinyaritha sub-catchment.	Level of awareness Level of alertness Willingness to pay for solar energy	Data triangulation models and PESTELI analysis, Likert scale analysis.
To evaluate the cost benefit ratios of different models of solar financing in Ngaciuma-Kinyaritha sub-catchment.	Household energy expenditure Ability to pay Willingness to pay	Cost benefit analysis; discounting method PESTELI Analysis Scenario analysis

Source: Author (2012)

CHAPTER FOUR

RESULTS AND DISCUSSION

4.1 Introduction

This chapter presents the results and discuses the findings from the research conducted in Ngaciuma-Kinyaritha sub-catchment, based on the stated objectives of the study. The survey used questionnaires to extract information from selected households. It also undertook Focused Group Discussion (FGD) in Gitimene (Upper Zone), Gankere (Middle Zone), and Kiamwitari (Lower Zone) to extract qualitative insight into household energy use (Figure 3.2). Lastly, it made use of interview guides to extract information from solar energy vendors, district development planners and other relevant experts in the field.

With inference from the IWM approach to environmental management, this study details out socio-economic and environmental indicators that interplay to promote sustainable energy transition as pertains in the Ngaciuma-Kinyaritha sub-catchment. In this chapter, the energy use patterns in Ngaciuma-Kinyaritha sub-catchment are characterized and the cost benefit analysis for solar energy use or Solar Home Systems (SHS) presented as observed in the study area. It also presents three alternative scenarios of solar energy financing with the aid of PESTELI analysis. Lastly, it details out the willingness and ability to pay for solar energy as well as the level of environmental awareness and alertness with respect to energy use, as indicators of the potential environmental significance of energy transition in the sub-catchment.

4.2 Socio-Economic Characteristics of Respondents

In order to fully understand the sources of energy used, the reasons behind the figures and the possible paths of change which is the objective one of this study, the study paid attention to the socio-economic characteristics of the Ngaciuma Kinyaritha subcatchment. It laid emphasis of the household dynamics, the educational attainments, the occupational backgrounds of households and their relative income variations since these elements were are known to be the determinants of household development choices.

4.2.1 Highest educational attainment

The study focuses on household solar energy use as an emerging phenomenon that could potentially displace household reliance on environmentally degrading energy sources like paraffin and woodfuels. A household, in this study is taken to be all individuals, related or unrelated, who share the same dwelling and the same source of income. Out of the 100 respondents sampled, 48% were male-headed-households while 52% were female-headed households. The study perceived that where as the gender distribution did not have a significant effect on the energy use pattern, the education level influenced the type and amount of energy consumed by a household. The highest education attainment by households as presented in figure 4.1 indicates that the majority (51%) of households had attained secondary school education while basic, tertiary and vocational education were 37%, 4% and 7% respectively.

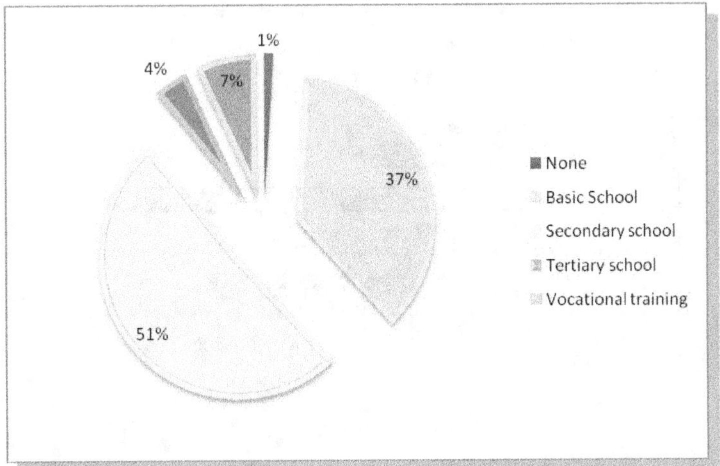

Figure 4.1 Highest educational attainments of households
Source: Author (2012)

4.2.2 Main occupation of households

Another household feature that influenced the pattern of energy use in the sub-catchment is the main occupation of the household or the livelihood options. The major livelihood option in the study area being agriculture (45%) included food crop cultivation such as maize and banana as well as cash crops like tea and Khat (Figure 4.2).

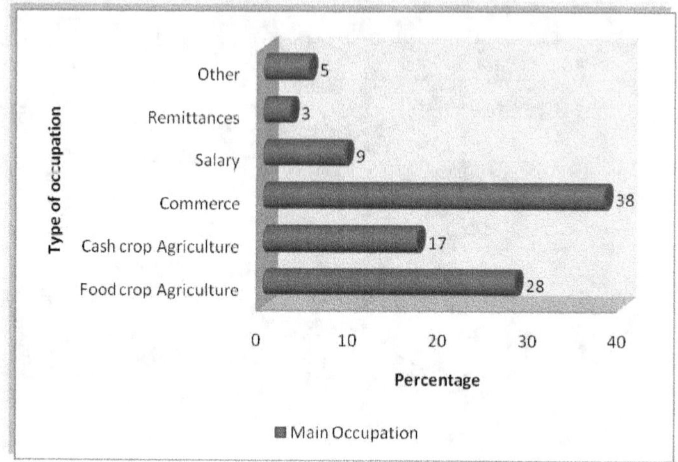

Figure 4.2 Occupation Distributions of Households
Source: Author (2012)

Commerce was found to be the second dominant occupation of households in the sub-catchment. The study perceived that whereas the occupation was not necessarily dependent on the education level, the income of the household was dictated by its educational background; as well as the size of the household (Figures 4.2 and 4.3).

4.2.3 Household income description

The study described the income status in the sub-catchment by matching the monthly income ranges with observed household size (Table 4.1). As indicated by GENI (2008), household income is a major determinant of energy access and use. It was noted that on the average, a household in the sub-catchment earned about 1,500ksh per month, with the highest limit being 21,000ksh and lowest being 500ksh. This is with regard to incomes in only monetary terms. Also, as illustrated in figure 4.1, the largest income groups in the study area range between 1,000Ksh to 5000Ksh (38%) followed by 10,000Ksh to 15000Ksh (34%).

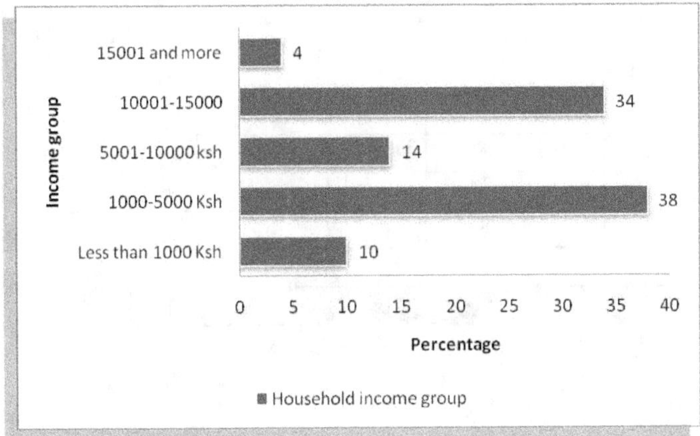

Figure 4.3 Household distributions by income group
Source: Author (2012)

It was noted that these income levels did not only determine the type of energy used by households. Out of the 17% households which spent more than 400Ksh on energy for cooking, 41% were high income households earning above 10000Ksh per month. The income status therefore also influences the quantity of energy used per period, the energy mix and the energy transition ability of each household. This makes it imperative to bring to perspective the income groupings by household size as shown in Table 4.1., which also gives a snapshot of the household economy in the study area.

Table 4. 1 Average Income per Month by Household Size Cross Tabulation

		Household size						Total
		1.00	2.00	3.00	4.00	5.00	6.00	
Average income per month	Less than 1000 Ksh	2	5	0	2	1	0	10
	1000-5000 Ksh	8	11	7	8	3	1	38
	5001-10000 ksh	3	2	3	3	2	1	14
	10001-15000	1	7	11	10	2	3	34
	15001 and more	0	0	2	1	0	1	4
Total		14	25	23	24	8	6	100%

Source: Author (2012)

For households who use the traditional stone fire, woodfuels is seen as a traditionally source of energy. In this regard, even with access to alternative energy sources, these households would not fully transform from the reliance on woodfuels. This transition inertia is also attributable to the perception that woodfuel is a free resource, which it means does not involve cash expenditure. However, according to ESDA (2003) the increasing scarcity of woodfuels in rural communities is increasing the social cost associated with fetching the resource. This study confirms the observation of ESDA, 2003. Conventionally, households in Ngaciuma-Kinyaritha collect woodfuel from cultivated farm lands, belonging to the household or to another individual. As a coping mechanism to the increasing woodfuel scarcity in the sub-catchment, most households have resorted to pruning the fresh trees in the *shamba* and drying it as firewood. Obando (2005), alluding to the IWM concept highlighted the interrelationship between vegetation cover and soil water or underground flow. Premised this interrelationship, the study perceived that the pattern of woodfuels in Ngacium-Kinyaritha has a negative implications for plant growth, forest resources, and agricultural productivity.

4.3 Characteristics of energy use in Ngaciuma-Kinyaritha sub-catchment

In order to evaluate household energy use the study characterizes energy use based on the source of energy, energy application, the cost associated, the reliability and the potential environmental effects associated with every form of energy used to meet daily household needs.

4.3.1 Sources of Energy Use in Study Area

From the Upper to the lower zones, it was noticed that the source of energy used is mainly dependent on availability. The most available source of energy in a given community is the most used. In Ngaciuma-Kinyaritha sub-catchment, household energy mix include, fuelwood, charcoal, paraffin, PV solar home systems, grid electricity, touch lights, chargeable battery and other biomass residue (Appendix 4.0). Appendix 4.0 shows the main sources of energy used for lighting in the sub-catchment. These include paraffin and PV solar home systems. The later is gradually increasing in the middle and the lower zones. The Naari zone (upper zone) was noticed to have less PV system preference due to low temperatures. Plates 4.1 and 4.2 are examples of energy forms used in Ngaciuma Kinyaritha sub-catchment.

Plate 4.1 Fuelwood collected from eucalytus trees in Murunguma
Source: Author (2012)

Plate 4.2 PV solar home system mountd on rooftop in Murunguma
Source: Author (2012)

Energy access is also differentiated based on income levels. However, it was noted that, while income levels determine more of the amount of energy consumed, the form of energy used is mostly determined by availability, both for lighting and for cooking. The study noted that the form of energy used for cooking has not in Ngaciuma-Kinyaritha has not transformed much since the 2009 population census study. However, energy use has transited due to the gradual shift away from the traditional stone fire to the improved traditional stone fire and from the ordinary Jiko to the improved Jiko (Figure 4.4). It was estimated that the improved traditional stone fire uses 20% less fuelwood than the traditional stone fire and the improved Jiko uses about 35% less charcoal than the ordinary Jiko. However, the transit to these improved cooking appliances has been slow with about 62% of households still sticking to the traditional stone fire, 22% using improved traditional stone fire and 5% using the improved Jiko (Figure 4.4). Figure 4.4 illustrates the household distribution of energy used for lighting and the major cooking appliance used.

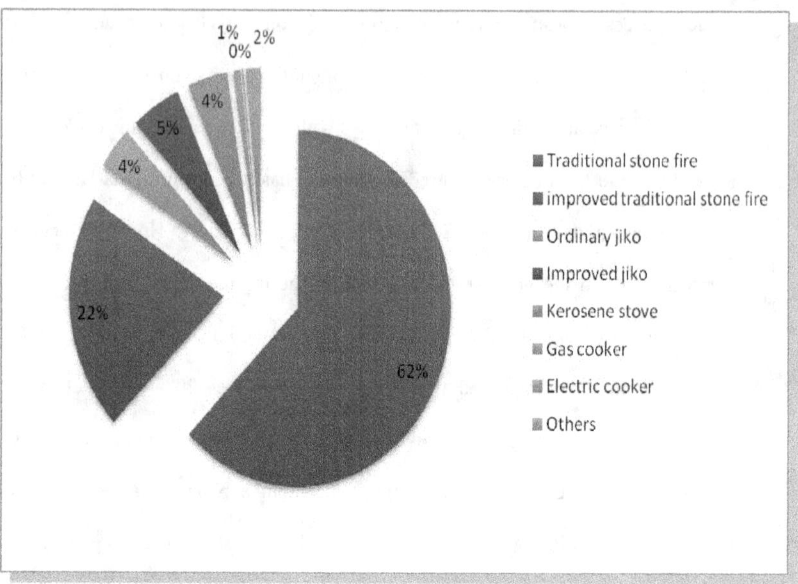

Figure 4.4 Distribution of energy use in study area
Source: Author (2012)

The improved Jiko is becoming more common due to the less smoke release and less wood consumption (Figure 4.4). Also, it was witnessed that most households use the traditional stone stove because of the type of meals they cook. A Household Head remarked that "if we use Jiko to cook *Githeri* or *mokimo*, consumes too much charcoal". Plates 4.3 and 4.4 are examples of traditional woodfuel use in Ngaciuma Kinyaritha sub-catchment. Some Households also see the traditional Stone stove to be faster in cooking than other appliances (Plates 4.1 and 4.2). Inferring from this observation, the study perceived that the switch from woodfuel to other energy forms, as observed by Mahiri, 2003, is different from the energy ladder concept, of switching from a superior quality fuel to an lower quality with increased scarcity, or from a low to a more technologically advanced fuel as the household income increases (Mahiri, 2003 pg 165).

Plate 4.3 A respondent heating water on a traditional Jiko at Munithi
Source: Author (2012)

Plate 4.4 A respondent attending to her traditional Jiko during an interview at Munithi
Source: Author (2012)

As indicated by Jacobson (2006), most rural communities in Kenya are deprived of electricity connection- Only 3.4% of electricity from the country's main electricity company, Kenya Power and Lighting Company (KPLC), is allocated to rural communities. Given that Ngaciuma-Kinyaritha is predominantly rural, deprivation is evident in the lower, middle and upper catchments. In order to give a broader picture of energy transformation. It was observed that about 64% of the communities prefer to use fuelwood, 30% prefer to use charcoal and the remaining 6% prefer other sources (Electricity, paraffin, Solar Panels,) for cooking and heating. For household lighting, about 20% prefer solar energy, 60% prefer the use of paraffin, and 20% prefer Electricity.

"If we get the solar energy that can be used for all our household needs, we can even extend it to our family members and for a small business like salon, Kinyonzi or even cold store and " ***(Respondent Household Head- Munithi community)***

This preference is also hinged to the cost associations of each source and point toward the willingness of households to shift away from the reliance on woodfuels. With regard to cooking, solar energy was seen not to possess the required capacity for cooking or heating activity at the household level. Instead, solar is gaining momentum in the aspect of household lighting, with increasing installations - about 22% of households using PV solar home systems.

4.3.2 Applications of Electricity in Ngaciuma-Kinyaritha

With the price of PV solar home systems decreasing over the years, recent installations tend to be relatively of higher capacity than the preceding years (Jacobson, 2006). The study perceived that the application of solar energy at the household level is also widening to include the use of 16inch Television set, Medium size refrigerator and high-end cassette players (Table 4.2). However, a higher capacity panel would be more useful for other appliances like a refrigerator, irrigation pumps and for running enterprises like hair saloon, barbering shops and others. Table 4.2 is the list of electricity consumption for commonly used household appliances used in Ngaciuma-Kinyaritha, indicating the capacities of PV SHS currently in use. Whereas grid electricity is been used for all household energy needs, PV SHS was only applicable to appliances that require less than 100watts. Thus, appliances including high end refrigerators and electric irons could not be used with PV SHS (Table 4.2).

Table 4.2 Type of Electric Appliance and Their Use

Appliance	Electricity use (watts)	Use in Grid Electricity	Use in PV systems
15 watt Incandescent light bulb	15	Often	Often
14 inch black and white television	10	Often	Often
Radio/cassette player	1	Often	Often
Mobile phone	3	Often	Often
14 inch colour television	70	Often	Often
Electric Iron	1,500	Sometimes	Not applicable
Small size refrigerator	80	Often	Sometimes
Electric Cooker/stove	1,500	Sometimes	Not applicable

Source: Author (2012)

4.3.3 Costs of Energy Use

As aforementioned, KPLC, the country's main electricity distributor, perceive rural energy supply to be less cost effective. As such it allocates 3.4% of its generated power to rural areas. Stemming from this poor connection, only 11 percent of households in Ngaciuma-Kinyaritha are connected. The study noticed that most rural households in Ngaciuma-Kinyaritha perceive electricity to be more expensive and unaffordable owing not to the monthly electricity bill but rather the capital requirements of connecting electricity to the house.

As one respondent indicated emphasized, *"I prefer to use solar energy in my house because it is cheap"*. In this statement, she referred to the initial capital requirements to connect to the national electricity grid. A typical 14watt solar panel in the subcatchment cost

17000Ksh in 2006. A household head who owned a solar panel explained that the cost of connecting electricity to her house as at then would have been about 25000Ksh. Plate 4.3 is an example of PV SHS installed on the roof top that is mainly used for lighting in the Ngaciuma Kinyaritha sub-catchment.

Plate 4.3 PV solar home system mountd on rooftop in Kauthini
Source: Author (2012)

A significant percentage, about 22 percent- mainly in the lower and middle zones, was noted to be using SHS household lighting owing to the reason given above (Plate 4.3). Table 4.3 shows the energy mix for household lighting, the initial capital requirements for access and the commonly used model of financing for each source. Apart from SHS, rural households were found to have access to a wide mix of energy technologies for household lighting (Table 4.3). The study noticed that the greater proportion (80%) of households had access to kerosene or paraffin while about 11% had access to grid electricity.

Table 4.3 Electricity Cost and Access Levels in Ngaciuma-Kinyaritha

Technology type	Population with access (percentage %)	Required Initial cost to access the Technology (in Kenya Shillings)	Average Monthly household expenditure per year (Ksh)	Common model of financing
Solar PV systems	22	17,000-50,000	58	Personal savings and loans
Rural grid connection	11	10,000-100,000	120	Personal savings and loans
Lead-Acid Battery systems	9	4,000-10,000	150	Personal savings
Dry cell battery	78	100-1000	50	Personal savings
Kerosene/ Paraffin	80	100-1000	150	Personal savings

Source: Author (2012)

Appendix 5.0 details out the Pearson Chi-Square tests for both hypotheses one and two of the study. Addressing the hypotheses (*1*), the field study indicated that a cost decrease in solar energy has a significant effect on the utilization of solar energy for cooking comparing (X^2=55.606, d.f=20, P=001), at a 95 percent confidence interval (Appendix 5.0). However, due to the high cost of electricity bills, even households that have electricity connection were observed to still prefer to use fuel woods because it is collected from the farms for free. Charcoal is the second dominant source that is more expensive. It was noticed that the charcoal used was not produced in the Ngaciuma-Kinyaritha. It was brought in from other parts of the Imenti-North District, thus relatively expensive with a bag costing about 850ksh.

4.4 Cost Benefit Analysis of PV Solar Home System (SHS)

About 90% of the PV panels that were in use had been acquired starting from the year 2000 and 15.4% were acquired between 2009-2012. This is an indication that the technology has left a measurable acceptance in the sub-catchment. This is reflected in the significant increased in the use of SHS from 6.6% in 2009 to 22% in 2012 (Table 4.3). As indicated by Jacobson (2004), this study confirms rather that solar energy is still being used largely for household lighting, powering the television; mostly black and white, the radio and for mobile phone charging (Table 4.2). Regarding cost of household energy for lighting (H_0), the study again noticed at a 95 percent confidence interval that a cost decrease in solar energy has a significant effect on the utilisation of solar energy at the household level (X^2= 34.138, df=20, P= 0.025) (Appendix 5.0).

4.4.1 Costs of a PV Solar Home System

Table 4.4 presents the cost element of acquiring and maintaining a SHS. The prevailing prices of PV SHS, sampled from three vendors in Meru town, Nairobi and other online vendors range from 12,000Ksh to 263,000Ksh (Table 4.4). The cost of installation includes, purchasing wires and sockets as well as workmanship. Maintenance is done usually after every three years. However, this study also included the cost of replacing bulbs as a maintenance item. It was observed that maintenance costs which is the only recurrent cost averages 1,000Ksh per year and an estimated 25 years is adopted as the life span of a PV SHS (Table 4.4). This recurrent solar energy cost, usually maintenance cost, includes the cost of acquiring a new battery after every three years, the cost of replacing light bulbs and replacing invertors. This maintenance cost was noticed to be affordable for most SHS users in the catchment based on the computed average household expenditure.

Table 4.4 Computing the Costs Photovoltaic (PV) Solar Home Systems (SHS)

PV System Capacity	Average System cost (Ksh)	Installation cost (Ksh)	Maintenance costs/annum (Ksh)	Total (Ksh)	Total cost in 5yrs (Ksh)	Total cost in 15yrs (Ksh)	Total cost in 25yrss (Ksh)
Less than 50 watts	20,000	2,000	500	22,500	25,000	30,000	35000
51-100 watts	50,000	3,000	2000	55,000	65,000	85,000	105,000
101-200 watts	80,000	5,000	2,000	87,000	97,000	117000	137,000
201-1Kilowatts	250,000	10,000	3,000	263,000	278,000	308,000	338,000

Source: Author (2012)

As noted in table 4.4 above, SHS costs as well as their respective maintenance costs tend to decrease as the capacity increases. For instance it observed from different vendors that whereas a standard 14watt PV model costs 17,000Ksh, a standard 50watts solar model costs 25,000 instead of 51,000. In much the same way, the average maintenance cost of higher capacity models also tend to decrease marginally from 500Ksh to 3000Ksh (Table 4.4). This is an opportunity that could be utilized to acquire higher capacity solar models for household use as well as small scale enterprises at the community level for household development.

4.4.2 Benefits of a Photovoltaic (PV) Solar Home System (SHS)

Table 4.5 presents the benefits of using SHS in Ngaciuma-Kinyaritha. The benefits included the avoided conventional energy cost, which is the total expenditure that households would have incurred on lighting without solar energy. To compute the avoided cost of lighting for households, the study noted the monthly average cost of paraffin, cost of connecting household to the national electricity grid and average monthly electricity bills per household. The details of these are shown in Appendix 7.0. The monetary value of avoided $C0_2$ was computed using the household emission factors given in table 3.2. In Kenya, the cost of carbon mitigation from solar electrification ranges from 9350Ksh to 11,900Ksh (Jacobson, 2007). This study used 11,900Ksh as cost of carbon mitigation due to solar energy. The monetary values of other social benefits such as returns on improved education, time savings for household chores and improved returns on home business were estimated with the aid of a guide provided by the World Bank (Table 4.5).

Table 4.5. Emission Factors by Type of Fuel

Benefit Category	Improved returns on education and wage income	Time savings for household chores	Improved productivity of home business
Benefit value(us US$/month)	37.07	24.50	34.00 (current business) 75.00 (new business)

Source: World Bank (2008)

Note: The shilling per dollar rate as at April 2012 is given as 85Ksh or US$1 (Central Bank of Kenya, 2012).

The study also computes these benefits for a period of 25 years using the discounting method as shown in Appendix 6.0. Leading among these benefit items is avoided conventional cost and avoided $C0_2$ emission. The study reveals that marginal benefit of solar energy tends to increase as the PV module capacity increases. This benefit attributable to the economic gains associated with higher capacity PV modules (Table 4.6). For instance, where as a 120watt model could be used to refrigerate a butchery and add to household income level, a 50watt model is limited to household lighting with no income gaining activity. Given this analogy, the study indicates that it's much easier for households to access higher capacity solar PV models for income generating activities than it is for lower-capacity non-income generating models.

Table 4.6 Computing the Benefits of Solar Home Systems (SHS)

PV System Capacity	Avoided conventional energy cost (p.a) (Ksh)	Avoided carbon dioxide emissions (p.a) (Ksh)	Time savings for household chores (p.a) (Ksh)	Returns on education & wage income (p.a) (Ksh)	Improved productivity of home business (p.a) (Ksh)	Total (Ksh)
Less than 50 watts	22,400	3,000	294	444.8	-	23,212.6
51-100 watts	32,400	3,500	500	600	-	33,614
101-200 watts	35,000	5,000	600	700	75	36,775
201-1Kilowatts	110,000	8,000	2,000	3,000	2,500	120,500

Source: Author (2012)

4.4.4 Payback Period from Net Present Value (NPV)

The cost benefit analysis used NPV of over an estimated lifespan of 25years to estimate the payback period as well as the life time payback of a SHS. It was noted that NPV values of all the selected models remain negative from year one to year five (Table 4.7). In year six, systems that are less than 50 watts begin to produce a positive NPV. Also, SHS that are more than 100watts show a positive NPV from year eight (after 8 years) (Appendix 6.0). Therefore, while the payback period for a 50watts PV model in Ngaciuma-Kinyaritha averages at six years, that of models that are more that 100watts capacity average at eight years. This payback period further accentuates the advantage of acquiring higher capacity models.

Table 4.7 Net Present Value (NPV) of Selected SHS in 5yrs

PV System Capacity	Total cost (Ksh)	Total benefit (Ksh)	$\frac{R_t}{(1+i)^t}$	NPV (Ksh)
Less than 50 watts	25,000	24,797	$\frac{-203}{(1+0.22)^5}$	(202.9)
51W-100W	65,000	51,323	$\frac{-13,677}{(1+0.22)^5}$	(13,670.2)
101W-200W	97,000	65,073	$\frac{-31,927}{(1+0.22)^5}$	(31,911)
201W-1kW	278,000	220, 950	$\frac{-169885}{(1+0.22)^5}$	(169800)

Source: Author (2012)

It was also noted that as the net payback of solar energy turns positive, it increases faster in higher solar models than in low capacity models. Echoing Jacobson's argument, where as is useful to pursue the neo-liberalists idea that solar energy market should be deregulated, it is more efficient advancing the Cory and Coughlin (2009) views on diversification. The study indicates that the NPV of higher capacity models tend to be marginally higher and environmentally more beneficial, in that, it could potentially off set the use of household reliance on national grid for ironing and for cooking which are both 1,500watt appliances (using a standard electric iron and electric cooker/stove).

The study indicates that a 1Kilowatt PV system accrues a total benefit 120,000Ksh instead of a proportionate 98,000Ksh in 5years averages above. Owing to this incremental marginal benefit associated with higher capacity systems, for instance a 1Kilowatts solar system, it is much beneficial economically, socially and environmentally to acquire higher capacity solar models that have the propensity to cut through the slow growth of the rural economy, especially in Ngaciuma-Kinyaritha.

Again, the study does not lose sight of the cost challenge. With the objective of emphasising sustainable financing models, the study highlights the neo-libralists idea of energy decentralization; including innovative models where the end user does not have to bear the full initial capital upfront as well as the periodic maintenance cost of a PV system. As indicated by Cory and Coughlin (2009), solar energy financing in recent years is taking a turn; whereby a third party is being introduced to procure and manage locally installed system to meet community energy needs. This can either be a joint community initiative or a Venture Capitalists (VC) who seek to make use of the government tax rebates on RETs to provide energy to rural and urban households.

4.4.5 Life Payback of a Photovoltaic (PV) Solar Home System (SHS)

The life payback of SHS is an indication of the profitability of household investment in the system. The life payback of is computed as the cumulated NPV of the system over its estimated life span (25years). Table 4.8 shows the cumulative benefits of selected SHS and their respective NPV over an estimated life span of 25 years. The NPV analysis below indicates a life payback of more than100,000Ksh on every 1 kilowatt system within a period of 25 years without government tax benefits. The cumulative NPV of a SHS, as computed, represent the life payback of the system; as indicated by the cost/benefit ratios (Table 4.8).

Table 4.8 Net Present value (NPV) of Selected SHS in 25yrs

PV System Capacity	Total cost (Ksh)	Total benefit (Ksh)	$\frac{R_t}{(1+i)^t}$	NPV (Ksh)
Less than 50 watts	35,000	43, 975	$\frac{8,975}{(1+0.22)^{25}}$	2,465.7
51-100 watts	105,000	136,613	$\frac{31,613}{(1+0.22)^{25}}$	8,684.9
101-200 watts	137,000	271,363	$\frac{134,363}{(1+0.22)^{25}}$	36,912.9
201-1Kilowatts	338,000	704,750	$\frac{366,750}{(1+0.22)^{25}}$	100,755.5

Cost Benefit Ratio

PV System Capacity	Total cost (Ksh)	Total benefit (Ksh)	B/C ratio $= \frac{\sum \frac{B_t}{(1+r)^t}}{\sum \frac{C_t}{(1+r)^t}}$
Less than 50 watts	35,000	43, 975	1.31
51-100 watts	105,000	136,613	1.30
101-200 watts	137,000	271,363	2.10
201-1Kilowatts	338,000	704,750	2.10

Source: Author (2012)

It is useful to note the change in the B/C ratio of the selected SHS in table 4.8. The B/C ratio refers to the benefit per cost of the SHS. The study indicates that whereas SHS with a capacity less than 100watts had B/C ratio of less than 1.5, SHS with more than 100watts had a B/C ratio of more than 2. This is means that for every one shilling expended on higher capacity systems, the return is two shillings. This again lends credit to the observation of Cory and Coughlin (2009) that financing higher capacity systems is emerging to be more viable than smaller SHS.

4.5 Cost Benefit Analysis of Solar cookers

Unlike solar PV systems, solar cookers have a lesser life span. For instance the standard Rudra Hot Box- Square or rectangle, with 3-4 people's meal-capacity, has a life span between eight to ten years. As observed in chapter two above, a standard solar box cooker cost an average of 20,000Ksh in Kenya. The maintenance and installation cost are regarded negligible since most of these can be done by the user. Table 4.8 summarizes the total cost and benefits associated with the use of solar cookers with a given life span of 9years.

Table 4.9 Cost Benefit Analysis of a Solar Cooker

System description	**Average System cost (Ksh)**	**Installation cost (Ksh)**	**Maintenance costs/annum (Ksh)**	**Total (Ksh)**
Rudra Hot Box- Square- 04 Product Capacity: Two hours cooking, 3-4 persons meal, Twice a day	20,000	-	-	20,000
Year	**Net Avoided conventional energy cost (70% of meals) (in shillings),**	**Avoided Carbone emission P.a (70%)**	**Total Benefits**	
Year one	2520	10,000		12,552

Year three	7,560	30,100		37,660
Year six	15,120	60,200		75,320
Year nine	22,680	90,300		112,980
Calculating the NPV				
Solar model	Total cost	Total benefit	$\frac{R_t}{(1+i)^t}$	NPV
Year one	20,000	12,552	-7448 (1+0.22)1	(6104)
Year three	20,000	37,660	17660 (1+0.22)3	9,703
Year six	20,000	75,320	55320 (1+0.22)6	16,415
Year nine	20,000	112,980	92980 (1+0.22)9	13,534

Source: Author (2012)

The net benefit of using solar cooker in this study include the net avoided conventional energy cost for cooking, avoided carbon emission cost and avoided health cost. A typical household in Ngaciuma-Kinyaritha spends 300Ksh on energy for cooking each month. This study observed that, about 70% of household cooking in Ngaciuma-Kinyaritha could be done with the use of solar cookers, thus displacing about 70% of energy expenditure (Table 4.9). Thus, in a typical year, average energy expenditure for cooking that could be displaced by a single solar cooking kit is 2520Ksh (Table 4.9). With a positive payback in the second year and a net payback of 13,534Ksh in year nine, the study noted that solar cooker is one of the most cost effective solar energy investments for small sized households.

4.6 Household Solar Energy Financing in Ngaciuma-Kinyaritha

The key stakeholders in the financing of household solar financing in Ngaciuma-Kinyaritha include: Individual Households, Equipment Vendors based in Meru, Financial institutions including Local SACCOs, Rural banks, Commercial banks, NGOs, Government Departments. Whereas the financial institutions, NGOs and government departments play a key role in providing the finance, the cost of solar energy is primarily determined by the Equipment vendors and Households with the major elements being system cost, cost of wiring and installation and maintenance cost.

Gueye *et al.* (2004) and Owusu (2010) had observed that the accessibility and use of solar energy vary significantly under different financing mechanisms. Following from that, different financing models and scenarios were analysed to determine the best-fit financing models. In line with that, this study further evaluated the conventional financing models as well as some unconventional innovations through interview guides and FGDs. The following financing models were evaluated: (a) Cash sales model, (b) third party ownership model, (c) joint community or group financing model and (d) third party ownership or solar developer model.

4.6.1 Cash Sales Model

In Ngaciuma-Kinyaritha, the cash sales model is observed to be the most common approach to financing solar energy dissemination. According to this model, the consumer purchases a PV system with some type of physical capital. "*The supplier may be contractually bound to provide technical assistance, although such an arrangement is not an aspect of this model*" (Gueye *et al.*, 2004). The study witnessed that about 85% of households using solar energy acquired it by cash purchases, about 85% had a systems capacity less than 50watts and that most of the solar energy systems had been in use for more than 4 years (60%). Table 4.11 presents the details of observed PV SHS in Ngaciuma-Kinyaritha.

Table 4.11 Capacity of a Typical Solar Home Systems (SHS) Module

PV System Capacity	Components	Appliances	Average Initial capital used (Ksh)	Percentage of households
Less than 20watts	PV solar Panel, Storage battery, wiring	3-4 bulbs	17,000	62%
20-50 watts	PV solar Panel, Storage battery, wiring	4-5 bulbs 1 black and white/colored TV set	25,000	36%
50-120 watts	PV solar Panel, Storage battery, wiring	6-12 bulbs 1 Colored TV set Small fridge	42,000	2%

Source: Author (2012)

It observed that less than 15% of solar energy users had acquired their model by other financing mechanisms such as salary loans, support from relatives living in Nairobi in the form of remittances and loans from relatives. However, these less than 15% was found to be relatively high capacity models ranging between 20watts and 50watts. Among rural energy users, it is well understood that this model alienates the rural poor from access to every form of energy, including solar energy. This model, in most cases, tends to promote the lowest-quality SHS on the rationale of cheap prices. This makes it imperative for the Energy Regulatory Commission (ERC) to introduce quality monitoring mechanisms in distribution of solar energy products. Nonetheless, the distribution of ability to pay for a PV SHS under the cash sale model was found to be relatively more varied with majority of households capable of saving about 200ksh per month towards the purchase (Figure 4.5).

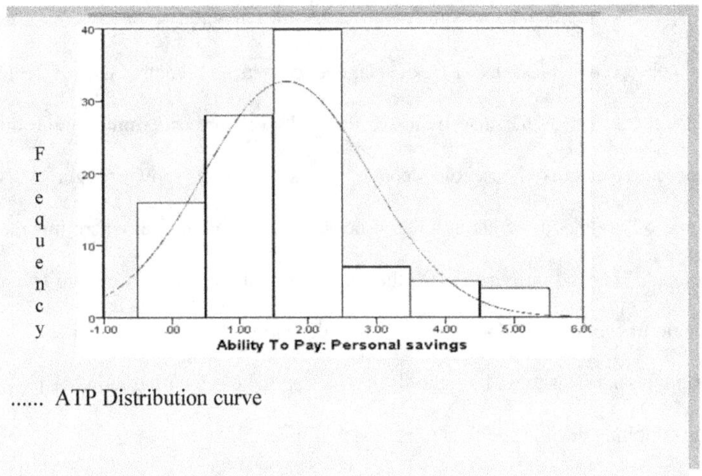

Figure 4.5 Ability To Pay (ATP) for SHS under cash sales model
Source: Author (2012)

4.6.2 Third-Party Credit Model

Third party credit is usually used to refer to financing model where consumers utilize end-user credit from a credit provider, such as micro-finance organizations, government agency or NGOs. In this study, third party credit is used to include dealer credits, salary loans and hire purchase agreements since they are all constraint by a common limiting factor and at the same time enhanced by similar factors.

The study witnessed with through FGDs that financial institutions hold the perception that financing SHS is not an income generating investment. Therefore, these institutions are reluctant about advancing credit to households for the purchase of SHS. As a result, the study observed that less than 10% of solar energy users had acquired their system by using salary loans. The FGDs also indicated that SHS vendors undermine option due to the unstructured nature of household incomes in the study are. As shown in figure 4.2, the main source of livelihood for about 45% of households is agriculture, where income flow is seasonal. The study observed also that, this seasonal income pattern tend to dictate the income flow of about 37% households who depend on commercial activities. This leaves the impression that these households may not be regular with the terms of payment and thus inhibiting the third party credits model of financing.

Figure 4.6 is an illustration of the energy expenditure at the household level per week. It also demonstrates the ability of households to pay for an alternative form of energy like PV SHS. As indicated, about 40% of households spend between 100ksh to 200ksh each week on energy for lighting while about 28% spend about 100ksh on energy for cooking (Figure 4.6).

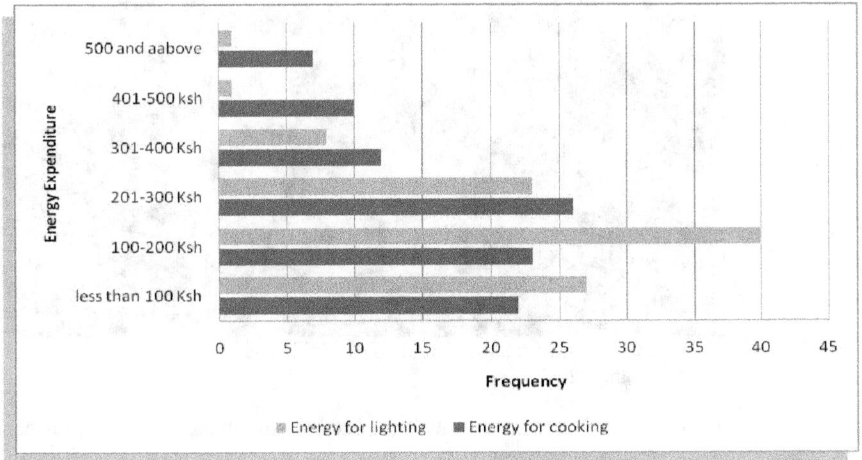

Figure 4.6 Household energy expenditure (for cooking and lighting)
Source: Author (2012)

Plate 4.5 Community members in a FGD at Githimene
Source: Author (2012)

The study witnessed from household income and expenditure analysis that on the average, each household would be able to contribute 200Ksh each month towards the purchase of a PV SHS. In FDG at Gitimene (upper zone), Gankere (middle zone), and Kiamwitari (lower zone), most households indicated a high preference of third party credit model as it effectively removes or reduces initial investment costs for SHS purchasing, allowing users to acquire higher capacity, high quality system (Plates 4.2 and 3.1). This potential is stifled by the seasonal nature of household income in the study area. As observed by Gueye *et al.* (2004); "*the lack of affordable credit schemes and/or other forms of financing mechanisms is regarded as the main barrier to widespread introduction of SHS in rural areas of the developing world*".

4.6.3 Third Party Ownership Model

Inasmuch as many studies (Gueye *et al.* 2004, Jacobson, 2006) have indicated that solar energy application in Kenya is limited to lighting, this study also aimed at exploring the other potentials of solar application in rural development. In view of this objective, it evaluates the cost of higher capacity solar models that have the potential to meeting all household energy needs including ironing, refrigerating and cooking- which can more sustainably be financed through third party ownership model. Not surprising though, the study records no third party ownership financing for solar energy in Kenya. However, it observes and describes the increasing potential of this financing model in contrast to Jacobson's observation in 2006.

Unlike the traditional solar PV model where a customer purchases a PV panel system for rooftop installation, the third-party ownership model is rapidly emerging as way for a households and institutions to deploy solar energy without providing up-front capital (Cory and Coughlin, 2009). Under this model, a solar PV developer installs, operates and maintains the system on behalf of the project owner and the host. An equity investor provides the upfront capital needed to the project and receives the benefits from the investment tax credits.

To facilitate third-party solar electricity production in Kenya, the MOE has created specific tax rebates for investors under the current policy document of Kenya, the vision 2030 (GoK, 2011). It avails a Feed-in-Tariff of 16.00Ksh per Kilowatt-hour for bulk solar energy producers in the first 20 years of operation. Given a growing middle class population, increasing rural demand for energy, and a decreasing per unit solar energy cost globally, the potential for bulk solar energy provision in Kenyan rural households is on the upsurge (Wamukonya *et al.*, 2002). Also, the ministry of energy indicates that even with the combination of geothermal, Hydro and wind energy supplies in Kenya, it is still challenging to be able to meet rural demand which constitute more than 70% of the population (MOE, 2009). Third party ownership in solar financing emerges therefore as a vibrant backstop leveraging on the widening capital market of which has incorporated PV SHS into its portfolio (Wamukonya *et al.*, 2002, Green, 2002)

4.6.4 Joint Community or Group Financing Model

The other financing model recently advocated for especially in rural areas is the Joint financing model. This may be community based or group based. The study witnessed that most households in Ngaciuma-Kinyaritha showed less preference for this model. Some of the reasons given includes, the difficulty in monitoring number of appliances used by each household, delays in payment and connection cost and likelihood of tribal alliances to creep into the management and render it inefficient.

4.7 Scenario Analysis of Solar Energy Financing With PESTELI

Although the establishment of fitting financing mechanisms is witnessed challenge to PV dissemination in Ngaciuma-Kinyaritha, there other elements that warrant significant consideration as Gueye *et al.* (2004) noted. These include economic stability, trade mechanism, financing mechanism and willingness to pay. This study brings into perspective the relevance of these factors through scenario analysis.

Scenario analysis has been used by public sector agencies, private sector organizations and non-Governmental organizations to manage risks predict future outcomes and develop strategic plans in the midst of uncertainties (Maack, 2001). Scenario analysis as a decision making tool in this study builds alternative financing environments and determines their respective rates of success using PESTELI (Table 4.12). It categorizes the observed macro environmental factors that enhancing or inhibit access and use of PV solar home systems into Political, economic, socio-cultural, technological, environmental, legal and institutional factors. It also presents specific strategies that could be used to harness the observed opportunities to the benefit of households while minimizing or eliminating the inhibiting factors.

Table 4.12 PESTELI Analysis of Solar Energy Financing

Indicator/Factor	Analysis
Political: P1: Strong communal Sense of initiative (e.g Naari and Thuraa water projects supply water for 32,000 households) P2: Observable levels of Corruption P3: De-regulation through the 1997 Electric Power Act and Electricity Act-2007	P1 and P5 can be used to minimize P2 at the community level. Also P5 can be used to facilitate P1 in accessing third party credits for solar home systems

P4: Removal of import, price, and foreign exchange controls P5: Reliable community cooperation (Naari and Thuraa water projects supply water for 32,000 households)	
Economic: Ec1: Availability of financial institutions (Meru Central SACCO, Equity Bank, Cooperative Bank) Ec2: Large energy deficit (3,000 MW as at 2010) Ec3: WEDI- improving the availability financial services for poor people in the rural areas Ec4: Rapid installation rates of solar modules (20% pa) Ec5: Decreasing capital cost of solar home systems Ec6: Relatively stable exchange rates Ec7: Relatively low propensity to save Ec8: Scaling-Up Renewable Energy Program (SREP).	WEDI has already disseminated 400 solar lamps to households. Leveraging on Ec5 and partnering with other institutions (Ec1), it could gain adequate capital to provide PV solar home systems. This would decrease Ec2
Socio-Cultural Sc1: Reliable community cooperation (Naari and Thuraa	It was realized that Sc4 serve a major contributing factor to Sc2 and Sc3. These together lead to poor access and use of solar energy. Solar energy

water projects supply water for 32,000 households) Sc2: Rapid rural-urban migration Sc3: Low education levels Sc4: Poor access to development infrastructure	promotion should therefore take into consideration Sc4 in order to achieve rural development
Technological: T1: Poor standardization T2: Rapid technology transfer T3: Increasing research and development T4: Decreasing capital cost of solar home systems T5: Decreasing maintenance cost T6: Increasing PV System capacities	En4, P3 and P4 could be used to enhance T2,T3,T4,T5 and T6. NEMA should also address issues relating to T1 with regard to solar energy dissemination
Environmental: En1: Poor access to development infrastructure En2: Low sunshine (Naari zone) En3: Scattered housing pattern (All three zones) En4: NEMA as a Designated National Authority (DNA) for the implementation of CDM in Kenya as required under the Kyoto Protocol.	En1, En2 and En3 are not supportive of the solar developer model and Joint community procurement. NEMA should consider these limitations in implementing CDMs

Legislative: L1: Difficulty in enforcing production standards standardization L2: De-regulation through the 1997 Electric Power Act and Electricity Act-2007 L3: NEMA as a Designated National Authority (DNA) for the implementation of CDM in Kenya as required under the Kyoto Protocol.	L2 has made it difficult to monitor L1 due to the influx of private manufacturers and distributors of solar energy products. NEMA as a DNA could develop mechanisms to reduce L1
Institutional: I1: The intensive cooperative management improvement system (ICMIS) I2: Availability of financial institutions (Meru central SACCO, equity bank, cooperative bank) I3: NEMA as a Designated National Authority (DNA) for the implementation of CDM in Kenya as required under the Kyoto Protocol.	I1 is a capacity building program. This could foster cooperation between NEMA and available financial institutions to promote CDMs

Source: Author (2012)

From the above analysis, it is worth indicating that the challenges associated with PV SHS dissemination are not necessarily insurmountable. With an appropriate community development framework, energy need analysis, sector goal harmonization and an efficient plan implementation management, access and use of PV SHS systems in Ngaciuma-Kinyaritha could be improved.

Jacobson (2006) had mentioned Global Environmental Facility (GEF) as a major stakeholder in the provision of funding for green energy which includes solar energy. In Kenya, the study observed that GEF has implemented and continue to implement a number of programs and projects to foster environmental conservation. Some of these projects include the National Capacity Needs Self-Assessment for Global Environmental Management (NCSA), Expedited Financing for (interim) Measures for Capacity Building in Priority Areas and Mount Kenya East Pilot Project for Natural Resource Management (MKEPP) which is running till the end of 2012. With regard to sourcing for financing for SHS in Ngaciuma-Kinyaritha, the identified niche is a simple goal harmonisation; creating a link between community development goals with regional and national development goals. The study perceives that this would avail significant financing either in the form of third party credits or third party ownership models for promoting access to SHS. Table 4.13 presents a continuation of the PESTELI analysis. It indicates the amount of influence exerted by each of the PESTELI factors as a basis for determining the central issues of intervention.

Table 4.13 Scenario Analysis Using PESTELI Factors

	Political factors	Economic factors	Socio-cultural factors	Technological factors	Environmental factors	Legal factors	Institutional Factors

Scenario one- Third party credits scenario	**P1, P2, P3, P4,**	**Ec1, Ec2, Ec3, Ec4, Ec5, Ec6,**	**S1, S2, S3,**	**T2, T3, T4, T5, T6**	**En1, En2, En3,.**	**L1, L2, L3**	**I1,I2, I3**
Scenario two- Joint community procurement scenario	**P1, P2, P3, P4,**	**Ec1, Ec2, Ec3, Ec4, Ec5, Ec6,**	**S1, S2, S3,**	**T1, T2, T3, T6**	**En1, En2, En3**	**L1, L2, L3**	**I1,I2, I3**
Scenario three- Third Party Ownership (Solar developer approach) Scenario	**P2, P3, P4, P5,**	**Ec1, Ec3, Ec4, Ec5, Ec6,**	**S1, S2, S3,**	**T1, T2, T3, T4, T5, T6**	**En1, En2, En3**	**L1, L2, L3**	**I1,I2, I3**

Source: Author (2012)

The level of influence exerted by a particular factor is denoted by difference in dept of the colour shade. The colour coding is determined by the number of issues under each factor that affect the scenario. From table 4.13, economic factors are found to have the greatest influence on all the scenarios. This supports the idea that initial capital cost is the major hindrance to solar energy access. Environmental factors have the lightest colour code in all three scenarios. This confirms the fact that the environment in Ngaciuma-Kinayritha largely supports the use of solar energy. This is further used to support the scenario analysis.

4.7.1 Third Party Credits Scenario

This scenario is based on the third part credits model. It assumes that micro-financing organizations and NGOs in Kenya would lend money to individual households in under agreeable terms purposely for the acquisition of solar home systems. One noted player in this scenario is Women Enterprise Development Institute (WEDI). WEDI operates as a micro-finance agent that promotes savings and lending activities among women's groups in the central province of Kenya where Ngaciuma-Kiyaritha is located. In partnership with Global Village Environment Program (GVEP), WEDI is already disseminated about 400 solar lamps for household lighting. It collects a deposit of 50% of the cost of the solar lamp (2100Ksh) from the households and the other 50% is paid in three equal instalments over a one year period.

This could be replicated by other micro-financing organizations Ngaciuma-Kinyaritha or scaled up by WEDI. Agreeably, it was unravelled in focused group discussions that inasmuch as households are pleased with third party credits, they would not be willing to accept it unless the terms of payment are well understood and are found to be convenient for them. The study also discovered that, under this scenario, 68% of households would be able to pay more than 200Ksh each month (figure 4.7). By paying 200Ksh every month, it would take an average household 21 months to acquire a 120watts SHS, costing 50,000Ksh.

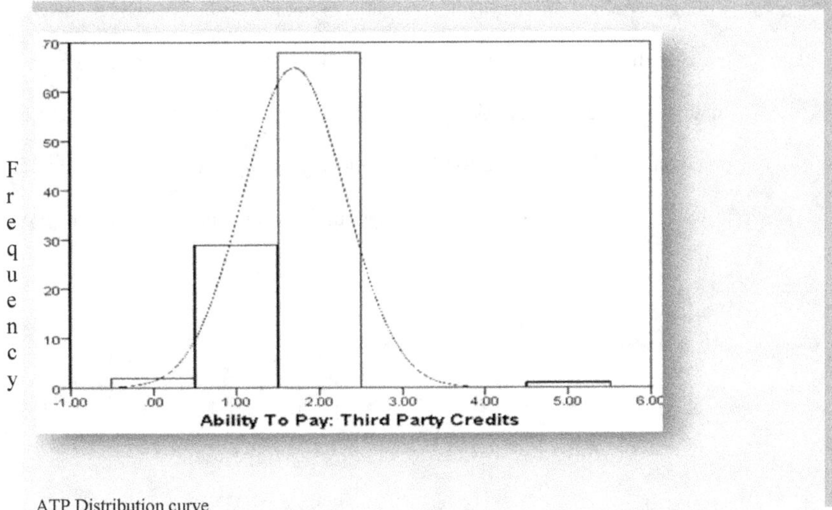

ATP Distribution curve

Figure 4.7 Ability to pay for SHS under third party credits model
Source: Author (2012)

It was also realized that with the third party credit scenario, households can improve their income and ability to pay by applying this solar energy in a business venture or enterprise and promote cottage industrialization in the sub-catchment. This would therefore require that the terms of repayment be revised periodically as the household income increases. The major challenge identifiable in this scenario is the capital availability on the part of the financiers and the lengthy legal and administrative processes required.

4.7.2 Joint Community Procurement Scenario

Joint community or group procurement is a non-conventional model in solar energy financing according to Cory and Coughlin, (2009). This model has been used in other development projects like community water services. Given the cooperation, measurable willingness and ability to pay, it is possible to implement this model replicate this model in solar financing. In this study, households rather expressed displeasure in joint community procurement. Less than 10% of households would be willing to undertake joint procurement of a solar home system. Through FGDs, households expressed the willingness to jointly procure a PV solar energy system for community development projects like schools and clinics. On the average, participants were willing to contribute about 50Ksh each month until the full cost is defrayed.

4.7.3 Third Party Ownership or Solar Developer Scenario

Unlike third party credits scenario, in the solar developer approach as described by Cory and Coughlin (2009), households do not have raise capital to purchase PV module. They pay for the electricity supplied from a distributor. This scenario assumes that; first, an investor installs a large capacity PV solar system in a local area and produce bulk solar electricity to households, takes advantage of government tax incentives. It also assumes that households would pay a monthly bill for electricity supplied from a locally installed system without owning the system.

Based on the PESTELI analysis (Table 4.13), third party ownership ranks more favourable among all the financing scenarios. The underlining factor to this was that, households could receive continuous energy supply to meet all household needs including cooking and heating, run a household enterprise, and power all household appliances. Secondly, it totally eliminates the initial capital cost which is the major setback to solar energy access and use.

As indicated above, the study disclosed that the average ability to pay (ATP) of households in Ngaciuma-Kinyaritha to be 200Ksh per month. It also observed a likelihood of households establishing rural enterprises and increasing their income levels if they had access to reliable energy. This is supported by the conducive environment revealed under the PESTELI analysis (Table 4.13). Against this background, the study disapproves the idea that it is not cost effective to connect rural communities to the national electricity grid. In view of the fact that about 74 percent of the population of Kenya is rural, the study perceives a tenable potential for rural electricity supply.

Also, Jacobson observed in 2006 that the model is not feasible in developing countries like Kenya. The biggest challenge being poor availability of investor capital and that investors are more interested in conventional energy sources like hydro and geothermal and wind energy. As indicated by the PESTELI analysis (Table 4.13), given the increasing capital market of Kenya and the incorporation of the solar energy into the countries energy mix, there exist a considerable potential to implement the third party ownership financing model in solar energy financing.

4.8 Potential Environmental Effect of Woodfuel Use

Kenya's energy mix can largely be considered as "clean"; with less environmental effect - petroleum, H.E.P, geothermal, wind, PV solar energy and woodfuels (MOE, 2010). However, this energy mix is highly differentiated among different clusters within the population, for instance, the rural and the urban populations (GENI, 2008). In the case of Ngaciuma-Kinyaritha, especially in the upper foodcrop growing zone, energy mix is highly dependent on woodfuels and paraffin.

The study noted in an FGD that the level of preference for solar energy was less than the middle and lower zones and this was attributed to the low temperature in the upper Naari zone, Maximum; 22^0 and minimum; 11^0 . The reliance on woodfuels and paraffin has a multiple environmental effect. First, the abstraction of wood and forest resources reduces the vegetative cover and exposes the land to high erosion and higher evaporation (Obando, 2005). This reduces the soil water availability for plant growth. The loss of vegetation cover also means less carbon sinks increasing GHG effect. Agriculture dependent communities like Ngaciuma-Kinyaritha are the most prone to the above mentioned environmental vulnerabilities owing to their. In the study area, though most people (55%) did not have adequate knowledge regarding the range of effects of energy use, they had the general awareness regarding the effect that energy use has on the environment (Table 4.14).

Using FGDs, it was noted that about 90% agreed to observable environmental effects of charcoal burning in the catchment. This understanding is of relevance to the environmental consciousness of energy use which is primal to sustainable energy use. In the final analysis, the consciousness of individuals or households is thought to be the most important and elementary. Thus, once this consciousness is raised, it creates a platform for sustainable choices of energy use in a given watershed (Kaneda, 2001).

4.8.1 Estimating Environmental Consciousness of Energy Choice

Environmental consciousness of energy use as in the literature review encapsulates the environmental knowledge, awareness, alertness and willingness to undertake environmentally proactive behaviours. These parameters were fed into a graded scale or likert scale (Table 4.14) to obtain interval level estimates of environmental consciousness in the study area. The five level intervals used are color coded from Zero environmental consciousness to high environmental consciousness. In this scale, a household was considered as environmentally aware if that household notices the environmental effects of woodfuel use or the benefits of using alternative forms of energy such as PV SHS. If a household is well informed about the effects of woodfuel use and measures that could be taken to reduce it, such a household is described as having the knowledge of environmental consciousness. Households that had taken any action, such as reduce the amount of energy consumed or purchased a more energy efficient cooking appliance, to preserve the environment were considered to be environmentally alert while households that are willing to take change or take certain actions in the form of energy transition or transformation to protect the environment were considered environmentally proactive.

Table 4.14 Likert Scale of Environmental Consciousness Of Energy Use

Interval scale / **Elements/Variables**	**Not/zero**	**Less**	**Not sure**	**Strongly**	**Very Strongly**	**Total Score**	**Rank**
Awareness level (1)	0	2	2	18	8	30	Third
	0%	(7%)	(7%)	(60%)	(26%)		
Knowledge in environmental effect of energy use (2)	0	4	16	12	4	36	Second
	0%	(11%)	(44%)	(34%)	(11%)		
Environmental Alertness in energy choice	4	3	4	0	0	9	Fourth
	(44%)	(33%)	(22%)	0	0		

(3)							
Willingness to be proactive (4)	4	4	8	20	8	44	First
	(9%)	(9%)	(18%)	(45%)	(18%)		

Source: Author (2012)

The study indicates a strong awareness (60%) of the environmental effects of overreliance on woodfuels. It also shows a relatively low environmental knowledge of woodfuel use. As shown in the likert scale above, the willingness to undertake proactive environmental actions is relatively high in Ngaciuma-Kinyaritha sub-catchment yet less than 1% of the population actually exercises any environmental alertness in their choice of energy. The study also perceived that about 11% were well informed about the full effects of woodfuel use including $C0_2$ emission, loss of vegetation, poor agricultural productivity and exposure to heart related diseases (Table 4.13). Also about 44% have a moderate knowledge regarding the full environmental effect of energy use. This is supported in a chi square analysis at a 95 percent confidence interval (X2= 49, df = 63 P= 0.91), thus failing to rejecting the null hypothesis that a increased access to solar energy does not have a significant effect on the environment (Appendix 5.0).

The increasing use of solar energy is an indication that households appreciate solar energy Technology, apart from its environmental potential. However, the study observed a high level of consciousness with about 45 percent of respondents were willing to move from woodfuel use to solar energy given adequate capacity and conducive financing mechanisms.

4.9 Summary of Study Results and Analysis

This study results and analysis as presented in this chapter dealt with the observations on the field regarding the forms or energy used to meet household energy needs, the benefit/cost analysis of PV SHSs and the different models of financing. In summary, the study perceived that energy for cooking and heating in the Ngaciuma-Kinyaritha sub-catchment did not vary across the three zones as did energy for lighting. It also noticed that, where as the dominant financing model for PV SHS is currently cash sale, third part credits and third party ownership models of financing offer a better opportunity to households to acquire higher capacity PV SHS that could be used for multiple functions including household enterprise. This means the two models have the potential to promote rural businesses and rural development in general. Throughout the sub-catchment, from the upper to the lower zone, environmental consciousness at the household level was observed to be generally low (Table 4.13). The study indicates that given an enabling environment, the sub catchment is ready to upload the full potential of solar energy to meet all their energy needs.

CHAPTER FIVE

SUMMARY OF FINDINGS, CONCLUSIONS AND RECOMMENDATIONS

5.1 Introduction

This study dealt with evaluating solar energy financing models for environmental conservation in Ngaciuma-Kinyaritha sub-catchment. To this end, it set the primary objective of analysing the cost effectiveness of different financing models for solar energy and to assess the relative environmental significance of these financing models. Its specific objectives were as follows:

- To characterize the different sources of energy used by households in Ngaciuma-Kinyaritha sub catchment.
- To assess the environmental consciousness of energy use by households in Ngaciuma-Kinyaritha sub catchment.
- To evaluate the cost benefit ratios of different models of solar financing in Ngaciuma-Kinyaritha sub catchment.

Practical survey instruments used in the field study included: Structured questionnaires, interview guides and focused group discussions (FGDs). Descriptive statistics used include frequencies, percentages and means, supported by chi square, likert scale and PESTELI analysis. A qualitative triangulation of methods and results was done to provide a better understanding of the inhibiting and enhancing factors under study. These helped in answering the underlying questions of the research which are:

- How are household energy needs met in Ngaciuma-Kinyaritha sub catchment?
- What is the level of environmental consciousness of energy use by households in Ngaciuma-Kinyaritha sub catchment?
- Which financing models are the most cost effective for solar energy in Ngaciuma-Kinyaritha sub catchment?

5.2 Summary of Findings

From the upper to the lower zones of Ngaciuma-Kinyaritha sub-catchment, it was observed that the energy mix of a household depended mainly on availability including, fuelwood, charcoal, paraffin, PV solar home systems, grid electricity, touch lights, chargeable battery and other biomass residue (GENI, 2008). It was noted that there is a gradual transition toward the use of the improved traditional stone fire and the improved Jiko which save about 20% and 35% fuelwood in household coking. However, about 62% of households still stick to the traditional stone fire, 22% using improved traditional stone fire and 5% using the improved Jiko.

In order to give a broader representation of energy transformation in Ngaciuma-Kinyaritha, the study also perceived the energy preference by households, apart from the current form being used. It was observed that about 64% of the communities prefer to use fuelwood, 30% prefer to use charcoal and the remaining 6% prefer other sources (Electricity, paraffin, Solar Panels,) for cooking and heating. For household lighting, about 20% prefer solar energy, 60% prefer the use of paraffin, and 20% prefer Electricity. This preference is also hinged on the cost associated with each source and point toward the willingness of households to shift away from the reliance on woodfuels. With regard to cooking, solar energy is never used in any cooking or heating activity at the household level. Instead, solar is gaining momentum in the aspect of household lighting, with increasing installations (about 22% of households using PV solar home systems).

About 90% of the PV panels that were in use had been acquired starting from the year 2000, which means the technology has left a measurable impact in the sub-catchment. This study confirms that solar energy is still being used largely for household lighting, powering the television; mostly black and white, the radio and for mobile phone charging, as perceived by Jacobson (2004). The study reveals that marginal benefit of solar energy tends to increase as the PV module capacity increases. This benefit attributable to the economic gains associated with higher capacity PV modules. For instance, where as a 120watt model could be used to refrigerate a butchery and add to household income level, a 50watt model is limited to household lighting with no income gaining activity. Given this analogy, the study indicates that it's much more easier for households to access higher capacity solar PV models for income generating activities than it is for lower-capacity non-income generating models.

The study records no third party ownership financing for solar energy in Ngaciuma-Kinyaritha Sub-catchment. However, it observes and describes the increasing potential of this financing model in contrast to Jacobson's observation in 2006. The study perceived that about 85% of households using solar energy acquired it by cash purchases and that, most of the solar energy systems had been in use for more than 4 years (60%). Less than

15% of solar energy users had acquired their model by other financing mechanisms other than cash purchases. However, these less than 15% was found to be high capacity models ranging above 50 watts. As Gueye *et al.* (2004) argued, owing to the limited number of actors associated with the cash sales model (the supplier and the consumer), the transaction costs can be minimal and the model is relatively easy to adopt within varied economic contexts. Among rural energy users, it is well understood that this model alienates the rural poor from access to every form of energy, including solar energy. In terms of product quality, this model also tends to promote the cheapest and lowest-quality SHS due to poor product quality standards and high-initial investment costs.

In a PESTELI analysis, it was witnessed that one key player in the third party credits model is Women Enterprise Development Institute (WEDI) which operates as a micro-finance agent that promote savings and lending activities among women's groups in the central province of Kenya where Ngaciuma-Kiyaritha is located. In partnership with other well established institutions, WEDI could facilitate the provision of credits for the dissemination of PV home systems in Ngaciuma-Kinyaritha.

The study also established that under this scenario, 68% of households would be able to pay more than 200 Kenya shillings each month. By paying 200 shillings every month, it would take an average household 21 months to acquire a 120 watts model (costing 50,000 shillings). It was also realized that with the third party credit scenario, households can improve their income and ability to pay by applying this solar energy in a business venture or enterprise like, canteen services for schools in the community, Hair Salons or food processing enterprises. This would therefore require that the terms of repayment be revised periodically as the household income increases. The major challenge identifiable in this scenario is the capital availability on the part of the financiers and the lengthy legal and administrative processes required.

The study established that households in Ngaciuma-Kinyaritha disclosed an ability to pay up to the tune of 200Ksh per month, on the average. It also noted a likelihood of households establishing rural enterprises and increasing their income levels if they had access to reliable energy. Against this background, the study argues against the general perception that rural electricity supply is not cost effective. This perception by KPLC is not completely agreeable. Also, Jacobson noted in 2006 that the model is not feasible in developing countries like Kenya. The biggest challenge being poor availability of investor capital and that investors are more interested in conventional energy sources like hydro and geothermal and wind energy. As indicated by the PESTELI analysis, given the increasing capital market of Kenya and the incorporation of the solar energy into the countries energy mix, there exist a considerable potential to implement the third party ownership financing model in solar energy financing.

With reference to the environmental dimension of woodfuel use, the study witnessed in FGDs that about 90% of households agreed to the environmental effects of woodfuel use. The abstraction of wood reduces the vegetative cover and exposes the land to high erosion and higher evaporation, increase CO_2 in the atmosphere. Whereas 60% of households are strongly aware of the environmental effects of woodfuel consumption, less than 1% of the households actually exercise environmental alertness in their choice of energy.

5.3 Conclusions

This study interprets the economic, social and environmental significance of different financing models for PV solar energy at the household level in Ngaciuma-Kinyaritha sub-catchment, Kenya. In the final analysis, the study indicated that solar energy is not found to be used in cooking or heating activity at the household level in Ngaciuma-Kinyaritha. Instead, solar is gaining momentum in the aspect of household lighting, with increasing installations coming to about 22% of households. The study indicates that in the year six, systems that are less than 50 watts begin to produce a positive NPV. SHS that are more than 100watts show a positive NPV from year eight (after 8 years). This payback period further accentuates the advantage of acquiring higher capacity models. Though the cost puzzle remains unsolved, in a PESTELI analysis of different scenarios, it was realized that certain potentials and opportunities available in Ngaciuma-Kinyaritha could be used to leverage back the energy deficits it currently faces. These potentials include a strong communal sense of initiative (Naari and Thuraa water projects supply water for 32,000 households), WEDI partnering with other institutions to could gain adequate capital to provide PV solar home systems, and availability of financial institutions including Meru Central SACCO, Equity Bank, Cooperative Bank. Lastly, the integration of renewable energy into the Kenyan energy mix has been relatively poorly structured. This poor policy structuring is seen in the disconnection between environmental pillars of development and energy projects such that the bulk of Kenyan households are captured in the woodfuel dependent population. A sustainable energy transition in Kenya would therefore mean making the IWM approach more visible in the energy flagship projects of the Kenya Vision 2030.

5.4 Recommendations

i. Energy-user remodelling

The study recommends that energy producers and distributors like KPLC need to remodel the energy-user threshold in order to develop suitable energy mix that meets the needs of all users including rural catchments like Ngaciuma-Kinyaritha. It is useful to note that energy users have different needs, interests and values. For instance, the industrial urban dweller's demand for energy is different from that of the rural dweller. Within these groups there are also different groups like, the extremely poor, slum dwellers, farming communities, rural enterprise demand, the growing middle class and many more. As an observation in this study, energy user remodelling would provide insight into the unique energy packages that best suits each category. Energy-User remodelling could easily be done using the Energy Transition Model (ETM) which uses verified energy data to predict the different outcomes of alternative scenarios. Plate 5.1 provides a downloadable link for better understanding of how the ETM works. ETM is an internet based application that could allow energy planners and policy makers to create different future outcomes by adjusting different factors that could change over a period of time.

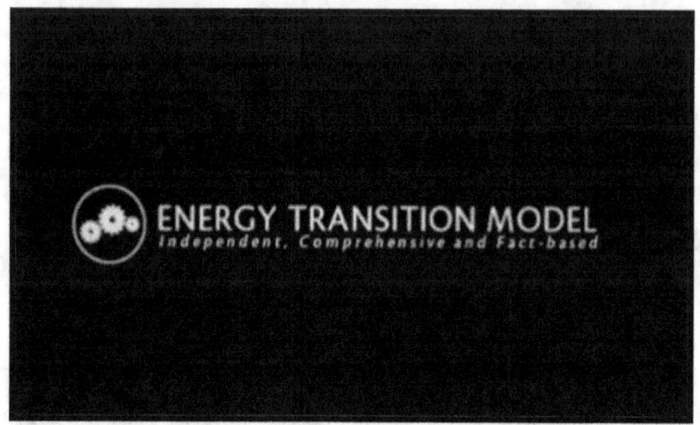

Plate 5.1 Demonstration of the Energy Transition Model

Source: http://et-model.com/pro, accessed on 14/04/2012.

ii. Energy Planning with LEAP

The study also recommends that the MOE builds its energy planning framework around the LEAP model (Long-range Energy Alternatives Planning model). As observed in this study, Kenya's energy mix is quite vast. Yet the rural population which constitutes the majority (over 70%) is highly deprived of energy for development. The study also indicated that there exists a reasonable market potential for rural energy supply. LEAP is a widely-used modelling tool for energy policy analysis and climate change mitigation assessment. It can also be used to track energy consumption and production as well as GHG emission sources and sinks. This model would help to monitor rural energy supply potentials in Ngaciuma-Kinyaritha and harmonise them with national energy development in order to reduce the continuous reliance on woodfuells and Paraffin.

iii. Integrated Capacity Building Approach

The study recommends that, it would be expedient to approach community capacity building from a more integrative approach. For instance, the WRUAs should not focus only on water issues but endeavour to build their capacities in accessing renewable energy for household needs. Rural households have been observed to have a relatively low propensity to change their energy use pattern due to the limited options available to them. As indicated in this study, different categories of energy users have different abilities to pay for energy used and this differentiates their energy access levels. To the extent that individual efforts have been commendable, local development organisations like WEDI, SACCOs and other NGOs should make an effort towards harmonising individual efforts and enhancing their opportunities to access solar energy.

iv. Piloting Alternative Financing Models for Solar Energy

Having identified and interpreted the significance of third party ownership financing model, this study recommends that it would be extremely beneficial for the MOE to begin to shift attention to solar electrification and creating partnerships to facilitate the production of bulk solar electricity in the rural parts of Kenya. The Ministry of energy is currently attempting to promote bulk solar electricity production in communities like Lamu, Lodwar, Mandera, Marsabit, Wajir and many more by giving tax incentives to prospective investors. However, much of the Ministry's efforts are currently slant towards wind energy, geothermal, small hydro dams, and nuclear energy. The difficulty is that despite this extensive energy mix, it is still challenging to reach the rural consumer and this does not only keep the energy deficit continuously high, it perpetuates the reliance on woodfuels by rural dwellers. The promotion of bulk solar electricity would spur cottage industrialization and rural development.

v. Promoting Higher Capacity SHS

It is further recommended that household access to higher capacity SHS needs to be improved. In this it perceives that local financing institutions like WEDI and NGOs need to create the leverage taking into account the ability to pay, which is 200Ksh per month. This may require structured collection arrangements agreed to by both parties. Access to higher capacity SHS is perceived in this study to have higher economic and environmental benefits to households and the sub-catchment in general.

vi. Piloting Solar Cookers in Sub-catchment

The study also recommends a piloting of solar cooking in the sub-catchment. As observed, the sub-catchment has shown appreciation of solar energy technology. Again, it would be the expedient for agencies like GVEP and GEF to pilot-study the acceptance and adoptability of various solar cooking technologies. This would facilitate the willingness of households to fully transit from woodfuels to solar energy technology.

vii. Promoting Environmental Consciousness in Energy Use

The study recommends that with the recent introduction of the county governance system, the Meru municipal planning department should design and implement integrated environmental consciousness-raising programs that would filter down the concept of environmental sustainability to the household level. This would include raising the household knowledge and awareness in environmental externalities of energy use and enhancing community alertness and willingness to undertake proactive measures that promote environmental sustainability. The study perceived that environmental consciousness could play a crucial role in the global outcry for an energy transformation. This would however require a sustainable political commitment to promote environmental sustainability at the local, provincial and national levels.

5.5 Further Areas of Research

This study mainly focused on evaluating the cost effectiveness of financing models for PV SHS and describing their environmental significance in Ngaciuma-Kinyaritha sub-catchment for the purpose of sustainable energy planning and sustainable energy use. For the purpose of comparison, this study could be replicated in a different catchment or on a larger geographic area different philosophical framework, data collection methods or different analytical tools. Other potential areas of research and development in Ngaciuma-Kinyaritha include:

- A comparative evaluation of the medium term impacts of wind and solar energy on the household economy
- An assessment of solar energy financing models for institutional and industrial needs
- Evaluating the potential of PV solar systems for rural enterprises

REFERENCES

Allen J. C. and Barnes D. F. (1985). The Causes of Deforestation in Developing Countries Annals of the Association of American Geographers, Vol. 75, No. 2, pg 174.

Alternative Energy Institute (n.d.). *Alternative Energy, Alternative Energy News And Information Resources About Renewable Energy Technologies* Http://Www.Altenergy.Org/Renewables/Hydroelectric.Html Date Accessed 12/01/2010.

Baumol, W.J. and Oates, W.E (1988). *The Theory of Environmental Policy*. Second Edition.Cambridge University Press. Cambridge.

Carraro M. and Massetti L. (2011). Beyond Copenhagen: a realistic climate policy in a fragmented world. Springer Science Business Media B.V. 2011

Chang J., Chen J., Shieh J. and Lai C. (2009) Optimal Tax Policy, Market Imperfections And Environmental Externalities In A Dynamic Optimizing Macro Model. *Journal of Public Economic Theory*, 11 (4), 2009, pp. 623–651, Wiley Periodicals, Inc.

Coase, R (1960), The Problem of Social Cost, *Journal of Law and Economics*, 3, 1-44.

Cory K. and Coughlin J., (2009). Solar Photovoltaic Financing: Residential Sector Deployment; National Renewable Energy Laboratory 1617 Cole Boulevard, Golden, Colorado www.nrel.gov

DAAD (2007). Summer School; Sub-catchment Management Plan- Ngaciuma-Kinyaritha Watershed, Kenya Nairobi, Kenya.

ESDA(2003). A New Energy Policy for Kenya? Policy Briefing No.1, Energy Alternatives AFRICA Ltd.

Fischer, G., F.N. Tubiello, H. van Velthuizen, and D.A. Wiberg, (2007). Climate change impacts on irrigation water requirements: Effects of mitigation, 1990-2080. *Technol. Forecasting Soc. Change*, 74, 1083-1107, doi:10.1016/j.techfore.2006.05.021.

Flyvbjerg, B. (2006). Five Misunderstandings about Case—Study Research. Qualitative Inquiry, *12* (2), 219—245.

Förch N. & Ngonzo C. (2009). Capacity Building In Integrated Watershed Management In Kenya An Independent Evaluation. A Conference Paper presented at the IDEAS Global Assembly; Johannesburg/ South Africa (18-20th of March 2009)

Gathenya M., Mwangi H., Coe R. and Sang J. (2011). Climate And Land Use-Induced Risks To Watershed Services In The Nyando River Basin, Kenya. *Jomo Kenyatta University Of Agriculture And Technology, Nairobi, Kenya and World Agroforestry Centre.*

GENI (2006). Overview of Renewable Energy Potential of India. Jawaharlal Nehru National Solar Mission; New Deli, India.

GENI (2008). *Proof It Can Exist: A Non-Subsidized Market For Photovoltaics In Rural Kenya-*
Http://Www.Geni.Org/Globalenergy/Research/Ruralelectrification/Casestudies/Kenya/Index.Shtml, accessed on 8th July, 2011.
Geothermal, Biogas and Solar Resource Generated Electricity, Nairobi, Kenya.

Gilg A., Barr S. And Ford N. (2005). Green Consumption Or Sustainable Lifestyles?
Government of Kenya (2011). Vision 2030, Ministry of Planning and national Development, Nairobi, Kenya.

Gollier C. (2009) Expected Net Present Value, Expected Net Future Value, and The Ramsey Rule, Toulouse School Of Economics (LERNA and IDEI), France.

Green M. (2002). *Solar Cookers: A Potential Mechanism for Challenging Gender Stereotypes.* Sustainable Development: An Oxymoron? pp. 62-67,
Agenda Feminist Media, Article Stable URL: http://www.jstor.org/stable/4066475

Gueye A., McNary J., and Okai J. (2004). Solar Energy and Rural Development: Constraints and Insights from the Developing World, http://www.gwu.edu/~oid/Solar_Energy.pdf, accessed on 8th July, 2011.

Hampel B. and Holdsworth R. (1996). Environmental Consciousness. *A Study in Six Victorian Secondary Schools*. Youth Research Centre. Faculty of Education, University of Melbourne, Parkville 3052

Hankins M. (2010). A case study on private provision of photovoltaic systems in Kenya. Energy Alternatives Africa (Kenya) Ltd, Nairobi Kenya, http://rru.worldbank.org/Documents/PapersLinks/27.pdf , accessed on 8th July, 2011.

Hilling D. (2011). Alternative Energy Sources for Africa: Potential and Prospects. Oxford journals 2011 pg 360-368.
Identifying The Sustainable Consumer, Department Of Geography, University Of Exeter, Exeter, UK.

IEA (2002). *Renewable Energy into the Mainstream*, The Novem Sittard, The Netherlands.

Imenti North District development plan, (2008). District Development Plan, Meru, Kenya.

IPCC (2000). Land Use, Land-Use Change, and Forestry. A Special Report of the IPCC. IPCC, Cambridge University Press, United Kingdom and New York, USA.

IPCC (2007). Chapter 3: Forestry, in Working Group III Report: Mitigation of Climate Change, IPCC Fourth Assessment Report, Cambridge University Press, New York.

Ishengoma, F.M. (2002). *"Modelling, Simulation and Digital Control of Photovoltaic power supply"* in a Summary of Dr.Ingeniør (PhD) Projects, at the Norwegian University of Science and Technology, NTNU, Faculty of Electrical Engineering and Telecommunications, Department of Electrical Power Engineering in 2001.

Jacobson A. and Kammen D.M. (2007) Engineering, institutions, and the public interest: Evaluating product quality in the Kenyan solar photovoltaics industry, Environmental Resources Engineering & Schatz Energy Research Center, Humboldt State University, USA and Energy and Resources Group & Goldman School of Public Policy, University of California, Berkeley USA, www.elsevier.com/locate/enpol

Jacobson E. A. (2004). Connective Power: Solar Electrification and Social Change in Kenya, Energy and Resources, Graduate Division, University of California, Berkerly.

Jacobson E. A. (2006). Field Performance Evaluation of Amorphous Silicon (a-Si) Photovoltaic Systems in Kenya: Methods and Measurements in Support of a Sustainable Commercial Solar Energy Industry, Humboldt University of Berlin, University Press, Germany.

Kaneda I. (2001). Environmental Globalism And Green Consumers, *Niigata Sangyo University, Japan*

Karekezi S. and Kithyoma W. (2003). Renewable Energy in Africa: Prospects and Limits African Energy Policy Research Network (AFREPREN)

Keen M., Brown, V. A., Dyball R., (2005). Social Learning in Environmental Management-Towards a Sustainable Future. Environmental Management and Development, Australia National University.

Kristoferson L. (2011). Energy and Environment in East Africa: Report from a Workshop. In: Royal Swedish Academy of Sciences, pg 1.

Kuhne T., (2005). What is a Model? Darmstadt University of Technology, Darmstadt, Germany

Kumekpor, T.K.B. (2002). *Research Methods And Techniques Of Social Research,* Accra, Sunlife Publications.

Lee P. (2007). 'Communication is peace: WACC's mission today', in *Media Development* 1/2007, pp. 49-53).

Maack, J. (2001). Scenario analysis: a tool for task managers. In: Social Development Paper no. 36. Social Analysis: Selected Tools and Techniques. World Bank, Washington, D.C.

MacKay, D.J.C. (2009). *Sustainable Energy- without the hot air,* England, UIT Cambridge Limited.

Mahiri I. O. (1998). The Environmental Knowledge Frontier: Transects With Experts And Villagers. Journal of International Development. Department of Geography, University of Durham, UK.

Mahiri I. O. (2003). Rural household responses to fuelwood scarcity in Nyando District, Kenya; Journal of International Development, Department of Geography, University of Durham, UK.

Ministry of Energy (2011).Feed-in-Tariffs Policy on Wind, Biomass, Small-Hydro, Ministry Of Energy (MOE), (2011). Scaling-Up Renewable Energy Program (SREP), SREP Investment Plan For Kenya, Nairobi, Kenya.

Murray B. C., Newell R. G., and Pizer W. A. (2009). Symposium: Alternative U.S. Climate Policy Instruments; Balancing Cost and Emissions Certainty. Oxford University Press, Oxford, UK.

Norman L. W. (2002). Depth of Knowledge Levels for Four Content Areas. http://www.nciea.org/publications/DOKreading_KH08.pdf

Obando J. A. (2005). Modeling Soil Erosion and Vegetation Change. FWU, Vol. 3, Topics of Integrated Watershed Management – Proceedings, pg 117-128, Department of Geography, Kenyatta University, Nairobi, Kenya.

Owusu A. (2010). Towards A Reliable And Sustainable Source Of Electricity For Micro And Small Scale Light Industries In The Kumasi Metropolis, College Of Architecture And Planning, KNUST, Kumasi, Ghana

Pigou, (1938). *The Economics Of Welfare* (fourth edition), London: Weidenfeld and Nicolson

Reddy S. (2008). Green Consumerism - Approaches And Country Experiences, Icfai University Press, Andhra Pradesh, India).

Sánchez M.J. and Lafuente R. (2010). Defining And Measuring Environmental Consciousness, University Pablo De Olavide. Sevilla. España, Revista Internacional De Sociología (RIS) Vol.68, N° 3, Septiembre-Diciembre.

Sankar U. (2005). Environmental Externalities. Madras School of Economics. Gandhi Mandapam Road, Chennai 600 025

Schulte-Bisping H., Bredemeier M., and Beese F. (1999). Global Availability of Wood and Energy Supply from Fuelwood and Charcoal. Ambio, Vol. 28, No. 7 (Nov., 1999), pp. 592-594. In: Royal Swedish Academy of SciencesStable URL: http://www.jstor.org/stable/4314963. Accessed: 02/07/2011 04:26

Sohngen B. (2008) Biofuels and Global Climate Change. AED Economics, Ohio State University, International Agricultural Trade Research Consortium, Annual General Meeting December 7-9, 2008, Scottsdale, Arizona, USA.

Stern N. 2007. *The Economics of Climate Change – The Stern Review*. Cambridge: Cambridge University Press, UK.

Tantawi P., O'Shaughnessy N., Gad K., Abdel M., and Ragheb S. (2009). Green Consciousness of Consumers in a Developing Country: A Study of Egyptian Consumers. Contemporary Management Research, Brunel University. Pages 29-50, Vol. 5.

UNEP (2006), Kenya: Integrated Assessment of the Energy Policy With focus on the transport and household energy sectors, Nairobi, Kenya.

UNEP (2010). Global Trends in Sustainable Energy Investment 2010; "*Analysis of Trends and Issues in the Financing of Renewable Energy and Energy Efficiency" http://www.rona.unep.org/documents/news/GlobalTrendsInSustainableEnergyInvestment2010_en_full.pdf, Accessed on 8/7/2011*

Vishnudas S., Savanije H.H.G. and Van Der Zaag P. (2005). " a conceptual framework for sustainable watershed management" In: *Proceedings of ICID 21st European regional Conference*, 15-19 May, 2005, Germany.

Wamukonya N., Martinot E., Chaurey A., Lew D., and Moreira J.R. (2002). Renewable Energy markets In Developing Countries; Global Environment Facility, 1818 H St. Nw, Washington DC.

Wilson E. B and Hilferty M. M (1931). The Distribution of Chi-Square. Department Of Vital Statistics, Harvard School Of Pujblic Health, USA.

World Bank (2008). Operational Guidance for World Bank Group Staff Designing Sustainable Off-Grid Rural Electrification Projects: Principles and Practices. The world Bank, Washington DC.

World Bank Group (2009). *Beyond Bonn: World Bank Group Progress on Renewable Energy and Energy Efficiency in Fiscal 2005–2009,* Washington, D.C. World Bank Institute, World Bank.

Youngquist W. (2000). *Alternative Energy Sources,* Http://Www.Oilcrisis.Com/Youngquist/Altenergy.Htm Date Accessed 02/02/2010.

APPENDICES

Appendix 1.0 Energy use in Imenti North District (Ngaciuma Kinyaritha subcatchment)

Households without electricity connection (No.)		56, 826
Trading centers connected with electricity (No.)		23
Trading centers not connected with electricity (No.)		43
Households using wood fuel		54, 531
Households using Kerosene		2, 850
Households using solar energy		126
Households using bio-gas		63
Households using improved woodfuel cooking stoves		886
Institutions (schools, hospitals, clinics, prisons etc) using improved woodfuel cooking stoves		3
Institutions (schools, hospitals, clinics, prisons etc) using LPG		12
HH distribution by main cooking fuel (%)		
	Firewood	86
	Paraffin	4.5
	Electricity	0.2
	LPG	0.8
	Charcoal	6.8
	Biomass residue	0.1
	Others	1.4
HH distribution by main lightening fuel		
	Firewood	2.0
	Grass	1.1
	Paraffin	76.8
	Electricity	12.6
	Solar	6.6
	Dry cell torch	0.6
	Candles	0.2
Household distribution by cooking appliance type		
	Traditional stone fire	62.4
	improved traditional stone fire	21.5

	Ordinary jiko	4.0
	Improved jiko	5.1
	Kerosene stove	4.2
	Gas cooker	0.8
	Electric cooker	0.2
	Others	1.6

Source: Imenti north District Development Plan, 2008

Appendix 2.0 Household questionnaire

DEPARTMENT OF GEOGRAPHY
SCHOOL OF PURE AND APLIED SCIENCES
KENYATTA UNIVERSITY
(Household Questionnaire)
Evaluating Solar Energy Financing Models for Environmental Conservation in Ngaciuma-Kinyaritha Sub-catchment

Name of enumerator:...
Date of interview:...
Start time:.......................................End time:..

Introduction

The information is required to enable the study identify the role of environmental consciousness in the choice of energy source. This information will enable the study examine how elicit community cooperation in sustainable energy consumption. Please, I assure you that any information provided would be treated with the deserving confidentiality and be used for purely academic purpose.
All the questions I have hear are related to this task and are open as possible. You may choose not to answer an item if it, in any way, has negative implications for you or your department.

Household Data (check key below)

1 Name	2. Age	3. Sex	4. Marital status					5. Highest Education attainment					6. Occupation					7. Employment status	8. Average monthly Income
			1	2	3	4	5	1	2	3	4	5	1	2	3	4	5		

	HH Energy Information (head of the household)			
Energy form	13 For Lightening	14 Cooking/heating	15 Type of Appliance Used	16 Quantities used per day
Firewood				
Paraffin				
Electricity				
LPG				
Charcoal				
Biomass residue				
Others				
Solar				
Dry cell torch				
Candles				
Others (Specify)...................				
17 Reasons for energy choice:				

Key for Energy Appliance type

Traditional stone fire	1
improved traditional stone fire	2
Ordinary jiko	3
Improved jiko	4
Kerosene stove	5
Gas cooker	6
Electric cooker	7
Others	8

Solar energy Use				
Type of solar energy system available		18. Known opportunities	19 Known challenges	
Off grid PV				
Gridded PV				
Off Grid CSP				

Gridded CSP				
What is the Capacity of your system (if applicable)?				
Less than 100 watts				
101-500 watts				
501-999 watts				
1000-1500 Watts				
More than 1500 watts				

Daily solar energy consumption				
Wood fuel use				
Charcoal:		20. Known opportunities	21. Known challenges	
Bags per week				
Amount spent per week				
		24. Known opportunities	25. Known challenges	
Fuel wood:				
22. Bundles/head loads per week				
23. Amount spent Per week				

Brief description of Household Economy

26. Income sources	27. Distribution (Amount)/ period	28. Reliability/ periodic variations
29. Main Expenditure items	30. In divisions of 10, what proportion goes to each of the major Items?	31. Level of recurrence

32. What savings method does your HH Use?

==

==

33. What is the major motivating factor for this savings?

==

==

34. Available sources of financing for solar energy in your community

Equipment cost	Amount spent (in shillings)	Payment Period (in months)
Cash purchases		
Loans from banks (give some details)		
Loans from NGOs (give some details)		
Other loans (give some details)		
Joint Purchases (give some details)		
Other sources (give some details)		
Accessories		
Means of financing	Amount (in shillings)	Payment Period (in months)
Cash payment		
Loans from banks (give some details)		
Loans from NGOs (give some details)		
Other loans (give some details)		
Joint Purchases (give some details)		
Other sources (give some details)		

35. Does your household use solar electricity?
Yes .... No ....

36. When was it acquired (Year of aquisition): ....

37. How was it financed (refer to item 34)?

==

==

38. Do you have any particular reason why you chose this financing means?
Yes .... No ....

<table>
<tr><td colspan="2">39. (If household uses solar energy) Has this affected the amount of woodfuel you use?
Yes .... No..........</td></tr>
<tr><td>40. By how much (in volume per day)?</td><td>41. By how much (in cash)</td></tr>
<tr><td></td><td></td></tr>
</table>

42. Have you ever had to maintain your system in any way? Yes .... No ....

43. What Kind of maintenance

44. How much on average do you spend?

45. How did you finance this?

46. Are there any specific needs you would like to use solar energy for (farm, industry, enterprise etc)

47. What capacity would you require for this (46 above)?

48. Which of the following models of delivery would you mostly prefer?

Solar models	Non Gridded Low capacity (100-500 watts)	Non gridded High capacity HSS (500-1000 watts)	Non Gridded industrial systems (1kw-10kw)	Grid tied system
Ligthenning				
Cooking/heating				
Production (farm, entreprice, industry etc)				
Other needs				

49. Which of the following models of financing would you mostly prefer?

Cash purchases	TP credits (Loans)		Group/community Joint financing		Solar Developer (Grid tied)
How much will you be willing to pay under the following financing models?	Per week (shillings)	Per month (shillings)	Per week (shillings)	Per month (shillings)	Per month (shillings)
	- 100	- 100	- 100	- 100	-50 shillings
	100-200	101-500	100-200	101-500	50-100
	201-300	501-1000	201-300	501-1000	101-150
	301-400	1001-1500	301-400	1001-1500	151-200
	401-500	1501-2000	401-500	1501-2000	201-250
	Above 500	Above 2000	Above 500	Above 2000	Above 250
51. Reasons for willingness level;					

51. Do you find any challenges associated with any of these models of financing:

---

52. Do you think that the use of firewood for cooking and heating affects the environment in any significant way?

---

53. How would you grade your level of awareness of environmental impact of energy use in your household?

Not	less		Moderately			Very		Very much	
0	0.1	0.2	0.3	0.4	0.5	0.6	0.7	0.8	0.9
0	0.1	0.2	0.3	0.4	0.5	0.6	0.7	0.8	0.9
Corresponding commitment to change									
Reasons for above choice: 1 2 3									

54. How did your household gain this awareness?

---

57. Under which of these five different provisions will your Household be more willing to use solar energy.

Model	Tally
Cash sales	
Third party credit,	
end-user credit,	

hire-purchase	
and fee for service	

58. Additional comments (*respondent*)
---------- ----------------- --------------------- --------------------------- -------------------------- ----------
------- -------------- ----------------- --------------------- --------------------- ----------------------------

59. Additional Observations (*Interviewer*)
---------- ----------------- --------------------- --------------------------- -------------------------- ----------
------- -------------- ----------------- --------------------- --------------------- ----------------------------

Appendix 3.0 Interview guides for Ministry of Energy and solar energy vendors

DEPARTMENT OF GEOGRAPHY
SCHOOL OF PURE AND APLIED SCIENCES
KENYATTA UNIVERSITY
Ministry of Energy - Renewable Energy Department

Evaluating Solar Energy Financing Models for Environmental Conservation in Ngaciuma-Kinyaritha Sub-catchment

Name of enumerator:...
Date of interview:...
Start time:..End time:..

Introduction
The information is required to enable the study identify the role of environmental consciousness in the choice of energy source. This information will enable the study examine how elicit community cooperation in sustainable energy consumption. Please, I assure you that any information provided would be treated with the deserving confidentiality and be used for purely academic purpose.
Your department is responsible for preparing the Ministry's budget proposals, incorporation of districts input and systematic scheduling of expenditures within the available resources. All the questions I have hear are related to this task and are open as possible. You may choose not to answer an item if it, in any way, has negative implications for you or your department.

1. Do you have specific financing policies for renewable energy sources in Kenya?

Yes:
………………………………………………………………………………………………………
………………………………………………………………………………………………………

No;
………………………………………………………………………………………………………
………………………………………………………………………………………………………

2. In which documents can these policies be found?

………………………………………………………………………………………………………
………………………………………………………………………………………………………
………………………………………………………………………………………………………
……

3. What are your targets for renewable energy production in Kenya?

………………………………………………………………………………………………………
………………………………………………………………………………………………………
………………………………………………………………………………………………………

4. Do you have separate budgetary allocations for promoting each individual renewable energy source in Kenya?

………………………………………………………………………………………………………
………………………………………………………………………………………………………
………………………………………………………………………………………………………

5. On the average how much is allocated for promote solar energy in Kenya?

………………………………………………………………………………………………………
………………………………………………………………………………………………………
………………………………………………………………………………………………………

6. Do you have any reservations regarding the amount allocated to finance solar promotion projects in Kenya?

………………………………………………………………………………………………………
………………………………………………………………………………………………………
………………………………………………………………………………………………………

The MOE has an attractive package for attracting bulk solar energy investors in which you give a substantial tax rebates for the first twenty years of operation. Could you tell me more about how this incentive is performing? Has some energy companies enlisted for this investment potential? If no, please give me your comment about this?

7. List companies (for above item here)

SN.	Name of company	Details of investment plans
Add notes:		

8. Do you have any reservations as to how solar energy in particular has been integrated into the Vison 2030 of Kenya?

....
....
....
....

9. Do you have any reservations as how solar energy could contribute to the development of Kenya?

....
....
....

10. Please, would you like to give a comment regarding how renewable energy is growing in the Kenya energy market?

....
....
....

DEPARTMENT OF GEOGRAPHY
SCHOOL OF PURE AND APLIED SCIENCES
KENYATTA UNIVERSITY
Interview Guide for Solar Energy Vendors

Evaluating Solar Energy Financing Models for Environmental Conservation in Ngaciuma-Kinyaritha Sub-catchment

Name of enumerator:...
Date of interview:..
Start time:...End time:...

Introduction

The information is required to enable the study identify the role of environmental consciousness in the choice of energy source. This information will enable the study examine how elicit community cooperation in sustainable energy consumption. Please, I assure you that any information provided would be treated with the deserving confidentiality and be used for purely academic purpose.
All the questions I have here are related to this task and are open as possible. You may choose not to answer an item if it, in any way, has negative implications for you or your department

1. *Interviewer*: Observe and list the solar modules

Solar module	Module capacity	Current price

2. Have you recently sold any module under the following purchase models?

Solar module	Cash sales	Third party credits	Group purchase	Solar developer

3. Would you like to give any general comments about solar energy prices, sales or purchases?

....
....
....
....

Appendix 4.0 Main energy for lighting

Energy Source	Frequency	Percent
Solar Electricity	22	22.0
Paraffin	65	65.0
Grid Electricity	11	11.0
Touch Light	1	1.0
Other	1	1.0
Total	100	100.0

Source: Field survey (2012)

Appendix 5.0 Pearson Chi-Square test for research hypotheses

Research Hypothesis:

1) The cost of solar energy does not limit accessibility by households in Ngaciuma-Kinyaritha sub-catchment.
2) Increased solar energy access by households has no effect on environmental degradation in Ngaciuma-Kinyaritha sub-catchment.

Chi-Square Tests Chi-Square Test for hypotheses one (1)			
	Value	Df	Asymp. Sig. (2-sided)
Pearson Chi-Square	34.138[a]	20	.025
Likelihood Ratio	33.393	20	.031
N of Valid Cases	100		

a. 23 cells (76.7%) have expected count less than 5. The minimum expected count is .01.

Chi-Square Test for hypotheses one (1)

	Value	Df	Asymp. Sig. (2-sided)
Pearson Chi-Square	55.606[a]	20	.000
Likelihood Ratio	32.115	20	.042
N of Valid Cases	100		

a. 22 cells (73.3%) have expected count less than 5. The minimum expected count is .07.

Chi-Square Tests for Hypotheses two (2)

	Value	Df	Asymp. Sig. (2-sided)
Pearson Chi-Square	49.244[a]	63	.898
Likelihood Ratio	35.217	63	.998
N of Valid Cases	100		

a. 82 cells (93.2%) have expected count less than 5. The minimum expected count is .01.

Appendix 6.0 Calculating CBR of selected PV SHS

	Avoided conventional energy cost p.a	Avoided carbon dioxide emissions pa	Time savings for household chores	Improved returns on education and wage income	Improved productivity of home business	Total	Total benefit 5th year	Total benefits 15 year	Total benefits in 25 year
Less than 50 watts	22400	US$3000	294	444.8	34	23212.6	38956	34,390	43, 975
51-100 watts	32400	3500	500	600	34	33614	51322.6	93967.8	136613
101-200 watts	35,000	5000	600	700	75	36,775	65072.5	133217.5	271362.5

201-1Kilowatts	110000	8000	2000	3000	2500	120,500	220,950	462,850	704,750

NPV of Selected PV SHS in 1st yr

Solar model	**Total cost**	**Total benefit**	$\frac{R_t}{(1+i)^t}$	**NPV**
Less than 50 watts	22,500	23212.6	$\frac{-203}{(1+0.22)^5}$	(202.9)
51-100 watts	55,000	33614	$\frac{-13,677}{(1+0.22)^5}$	(13,670.2)
101-200 watts	87,000	36,775	$\frac{-31,927}{(1+0.22)^5}$	(31911)
201-1Kilowatts	263,000	120,500	$\frac{-169885}{(1+0.22)^5}$	(169800)
		NPV of Selected PV SHS in 15yrs		
Solar model	**Total cost**	**Total benefit**	$\frac{R_t}{(1+i)^t}$	**NPV 1.37**
Less than 50 watts	30,000	34,390	4,390 (1+0.22)15	3204
51-100 watts	85,000	93967.8	8967 (1+0.22)15	6545.3
101-200 watts	117000	133218	16218 (1+0.22)15	11838.1
201-1Kilowatts	308,000	462,850	154,850 (1+0.22)15	113029.2

Calculating energy expenditure savings per household

1 headload of firewood = 10 Kgs or 4.5 pounds
Avearge weekly firewood use=15kg
Annual firewood use 15x4x12 = 720
1 Kg of firewood used = 1580g of $C0_2$ emission
$C0_2$ emission from firewood = 1137600/1000
$C0_2$ emission from firewood use per household = 1137.6 kg per annum
=1.1376 tonnes (metric)

Therefore;
At a rate of US$140 per ton of $C0_2$ from solar mitigation,
Avoided carbon emission cost per household from the use SHS = 1.1376x 140
= US$159 or 14333 Ksh

Avoided carbon emission cost per household in three years = 3x 1.1376x 140
= USD 478 or 43000 Kenya shillings

Avoided carbon emission cost per household in six years = 6 x 1.1376x 140
= US$956 or 86000 Kenya shillings

Avoided carbon emission cost per household in 9 years = 9 x 1.1376x 140
= USD1433 or 129000 Kenya shillings

Average quantity of paraffin used per month = 8 liters
Avoided emission from paraffin due to solar is given as 2.96Kg/ litre
Therefore;
Monthly $C0_2$ mitigation = 8x2.96
=23.68 kg/month
Annual $C0_2$ mitigation = 23.68x12
=284.16 kg
1kg = 0.001 tonnes
Therefore;
Annual $C0_2$ mitigation = 0.28kg
Cost of $C0_2$ mitigation = US$140/ton
0.28x140= 39.8
Average cost of $C0_2$ mitigation p.a= US$39.8 or 3,582Ksh

Appendix 7.0 Household Energy Use Characteristics

Main energy for lighting by Lighting energy cost per week Cross Tabulation								
		Lighting energy cost per week						Total
	Main energy for lighting	**less than 100 Ksh**	**100-200 Ksh**	**201-300 Ksh**	**301-400 Ksh**	**401-500 ksh**	**500 and above**	
	Solar Electricity	13	7	2	0	0	0	22
		5.9	8.8	5.1	1.8	.2	.2	22.0
	Paraffin	13	29	16	6	1	0	65
		17.6	26.0	15.0	5.2	.7	.7	65.0
	Grid Electricity	0	3	5	2	0	1	11
		3.0	4.4	2.5	.9	.1	.1	11.0
	Touch Light	1	0	0	0	0	0	1
		.3	.4	.2	.1	.0	.0	1.0
	Other	0	1	0	0	0	0	1
		.3	.4	.2	.1	.0	.0	1.0
Total		**27**	**40**	**23**	**8**	**1**	**1**	**100**

Average income per month by Cooking Energy Cost per week Cross Tabulation							
	Cooking Energy Cost per week					Total	
Average income per month	less than 100 Ksh	100-200 Ksh	201-300 Ksh	301-400 Ksh	401-500 ksh	500 and aabove	
Less than 1000 Ksh	5	3	2	0	0	0	**10**
1000-5000 Ksh	7	13	9	4	3	2	**38**
5001-10000 ksh	3	4	3	1	1	2	**14**
10001-15000	7	3	11	6	5	2	**34**
15001 and more	0	0	1	1	1	1	**4**

Total	22	23	26	12	10	7	**100**

www.ingramcontent.com/pod-product-compliance
Lightning Source LLC
Chambersburg PA
CBHW072305200526
45168CB00014B/508

* 9 7 8 1 5 1 4 2 6 8 5 0 6 *